Fundamentals of Plant Pathology

Fundamentals of Plant Pathology

Catlyn Bennett

SYRAWOOD
PUBLISHING HOUSE
New York

Published by Syrawood Publishing House,
750 Third Avenue, 9th Floor,
New York, NY 10017, USA
www.syrawoodpublishinghouse.com

Fundamentals of Plant Pathology
Catlyn Bennett

International Standard Book Number: 978-1-64740-065-1 (Hardback)

Cataloging-in-Publication Data

Fundamentals of plant pathology / Catlyn Bennett.
p. cm.
Includes bibliographical references and index.
ISBN 978-1-64740-065-1
1. Plant diseases. 2. Plants--Wounds and injuries. 3. Diseased plants. 4. Crop losses.
5. Seed pathology. I. Bennett, Catlyn.

SB601 .F86 2022
632.3--dc23

Table of Contents

Preface

This book has been written, keeping in view that students want more practical information. Thus, my aim has been to make it as comprehensive as possible for the readers. I would like to extend my thanks to my family and co-workers for their knowledge, support and encouragement all along.

The organisms which can cause diseases or infections are referred to as pathogens. The scientific study of diseases in plants that are caused by pathogens and various physiological factors is referred to as plant pathology. Some of the organisms that can cause infectious diseases are viruses, bacteria, protozoa, nematodes, phytoplasmas, fungi and oomycetes. Plant pathology deals with pathogen identification, disease cycles, disease etiology, disease resistance and plant disease epidemiology. The physiological plant disorders which are caused by natural processes such as frost, snow, drought and flooding are also studied within this field. The topics included in this book on plant pathology are of utmost significance and bound to provide incredible insights to readers. It brings forth some of the most innovative concepts and elucidates the unexplored aspects of this field. Those in search of information to further their knowledge will be greatly assisted by this book.

A brief description of the chapters is provided below for further understanding:

Chapter - What is Plant Pathology?

The scientific study of diseases in plants that are caused by pathogens and environmental conditions is referred to as plant pathology. It also deals with pathogen identification, disease cycles and pathosystem genetics. This is an introductory chapter which will introduce briefly all the significant aspects of plant pathology.

Chapter - Biotic Plant Pathogens

The biotic plant pathogens include various eukaryotic and prokaryotic organisms. Some of the common biotic plant pathogens are fungi, bacteria, virus, nematodes, mycoplasma, parasitic plants, magnaporthe grisea, virtoids, phytoplasma and oomycetes. This chapter discusses in detail these biotic plant pathogens.

Chapter - Abiotic Plant Pathogens

Abiotic plant pathogens include a number of soil-born plant pathogens such as phytophthora cinnamomi, gaeumannomyces graminis and agrobacterium tumefaciens. Other factors such as nutrient deficiency, extreme temperatures and toxic chemicals like herbicides, fungicides and air pollutants can also cause plant diseases. This chapter has been carefully written to provide an easy understanding of these abiotic plant pathogens.

Chapter - Common Plant Diseases

The different plant diseases are broadly classified into fungal, bacterial and viral plant diseases. Some of the fungal diseases are fungal leaf spots, rice blast fungus, botrytis cinerea, stem rust, etc. Aster yellows, bacterial wilt, bacterial spot, bacterial blight, etc. are some of the bacterial diseases. The most common viral diseases are phyllody and barley yellow dwarf. This chapter discusses all these plant diseases in detail.

Chapter - Diagnosis and Control of Plant Diseases

The diagnosis of plant diseases refers to the process of determining which disease is affecting the health of the plant. Some of the principles of plant disease control are avoidance, exclusion, protection and eradication. All these diverse principles related to diagnosis and control of plant diseases have been carefully analyzed in this chapter.

Catlyn Bennett

1

What is Plant Pathology?

The scientific study of diseases in plants that are caused by pathogens and environmental conditions is referred to as plant pathology. It also deals with pathogen identification, disease cycles and pathosystem genetics. This is an introductory chapter which will introduce briefly all the significant aspects of plant pathology.

PLANT DISEASE

A plant disease is defined as "anything that prevents a plant from performing to its maximum potential." This definition is broad and includes abiotic and biotic plant diseases.

Abiotic or Non-infectious Diseases

These diseases are caused by conditions external to the plant, not living agents. They cannot spread from plant to plant, but are very common and should be considered when assessing the health of any plant. Examples of abiotic diseases include nutritional deficiencies, soil compaction, salt injury, ice, and sun scorch.

Frost injury on soybean seedlings.

Biotic or Infectious Diseases

These diseases are caused by living organisms. They are called plant pathogens when they infect plants. Pathogens can spread from plant to plant and may infect all types of plant tissue including leaves, shoots, stems, crowns, roots, tubers, fruit, seeds and vascular tissues.

Soybean plants dying from Sclerotinia infection.

Types of Plant Pathogens

Plant pathogens are very similar to those that cause disease in humans and animals. Fungi, fungal-like organisms, bacteria, phytoplasmas, viruses, viroids, nematodes and parasitic higher plants are all plant pathogens.

Fungi and Fungal-like Organisms (FLOs)

Collectively, fungi and FLOs cause the most plant disease than any other group of plant pathogens. These organisms cannot make their own food, lack chlorophyll, have filamentous growth, and may or may not reproduce by spores. Fungi and FLOs are able to overwinter in soil or on plant debris. However, some fungi and FLOs cannot overwinter in northern climates because of low winter temperatures. These pathogens overwinter in southern climates and then are transported by air currents back to northern climates. Disease movement from southern to northern climates can be monitored during the growing season.

Soybean infected with Sclerotinia.

Bacteria

Bacteria are single-celled microscopic organisms with cell walls that reproduce by binary fission (one cell splits into two). Introduction to the plant must occur through natural openings or wounds in the plant. Bacteria overwinter primarily in soil and in or on plant material that does not decompose, but some survive inside insect vectors.

Soybean infected by bacterial blight.

Phytoplasmas

Phytoplasmasare microscopic, bacteria-like organisms that lack cell walls and thus appear filamentous.

Aster yellow phytoplasm infecting aster.

Viruses and Viroids

Viruses are intracellular (live inside the cell) nucleic acid particles with a protein coat that infect other living organisms and replicate in the hosts they infect. Viroids are virus-like particles but lack a protein coat. Viruses and viroids are primarily transmitted by vectors including insects, nematodes, and fungi, which introduce the virus or viroid during feeding. Viruses and viroids can also be transmitted through seed, vegetative propagation and pruning.

Soybean infected with Bean Pod Mottle Virus.

Nematodes

Nematodes are microscopic worm-like animals. The majority of nematodes are soil dwelling animals and move with soil. However, there are some nematodes that are transmitted through insects and infect above ground plant parts.

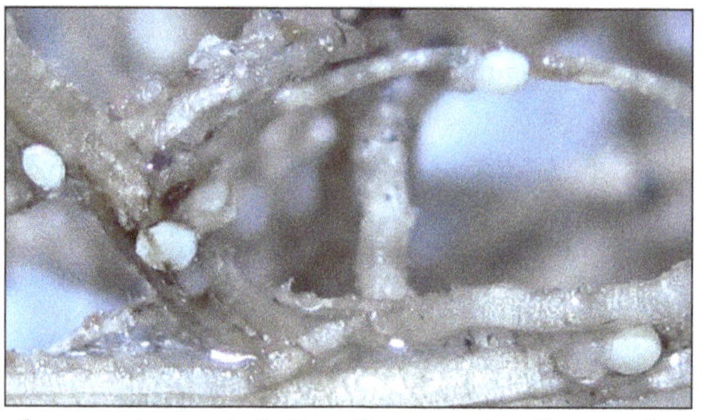

Adult soybean cyst nematode females emerging from soybean roots.

Parasitic Higher Plants

Parasitic high plants are plants that contain chlorophyll but cannot produce their own food. They parasitize other plants to obtain nutrients and water. Examples include mistletoe and dodder.

Disease Triangle

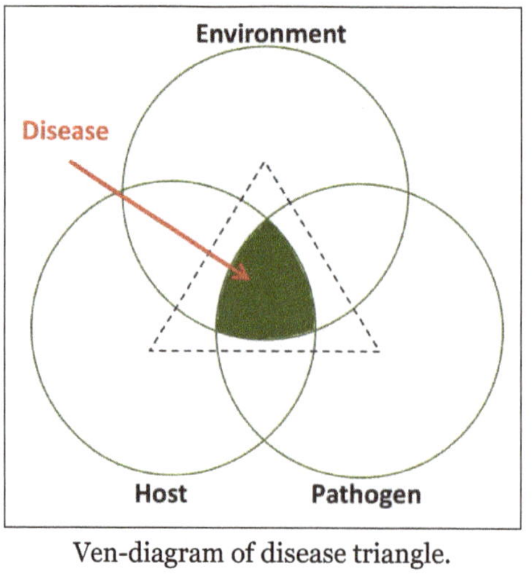

Ven-diagram of disease triangle.

Three components are absolutely necessary in order for a disease to occur in any plant system. The three components are:

- A susceptible host plant,
- A virulent pathogen,
- A favorable environment.

When these three components are present at the same time, a disease (shaded region) will occur if a susceptible host plant is in intimate association with a virulent plant pathogen under favorable environmental conditions. This concept is represented by the shaded portion of the diagram above. When there is a high degree of overlap (as the shaded area becomes larger), there will be a moderate to high amount of disease.

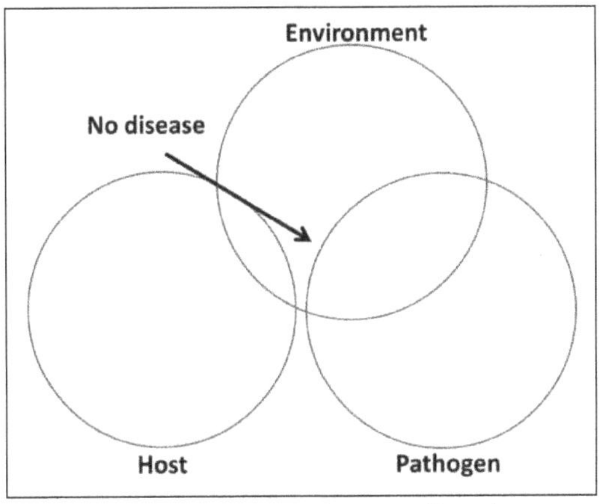

Variables within each component of the disease triangle may affect the presence of disease. This diagram represents a system in which the host is displaying resistance to disease even in intimate association with the pathogen under favorable environmental conditions.

It is important to remember that within each of the three components –host, pathogen, and environment –there are numerous variables that may affect both the incidence and severity of the disease. These variables include genetic diversity, biology and lifecycle of the host plant and pathogen, and environmental conditions.

- Genetic diversity: Within one species of host plant there may be an incredible range of genetic diversity that greatly influences susceptibility to any particular species of pathogen. If the host is resistant to a pathogen, even when the pathogen is present under favorable environmental conditions, diseases will not occur. Genetic diversity also plays a role in pathogen virulence or its ability to infect a host and cause disease, which may also influence the amount and severity of a disease.
- Biology and lifecycle of the host plant and pathogen: Host plants may be resistant to pathogens at one stage of development but not at another. In a similar manner, some pathogens must be at a critical life stage in order to cause infection.
- Environmental conditions: There are numerous variables in the environment that influence disease incidence and severity including temperature, sunlight, moisture, relative humidity, and time of year. Pathogens are typically restricted to an area based on the conditions of the macroclimate. A microclimate is the prevailing climatic conditions in a

certain geographical area. Within a macroclimate, small areas may exist in which the climate may be different than the surrounding areas. This is called a microclimate. Each landscape is filled with microclimates that exist because of differences in exposure to sun and wind, soil type and many other factors.

Disease Cycles

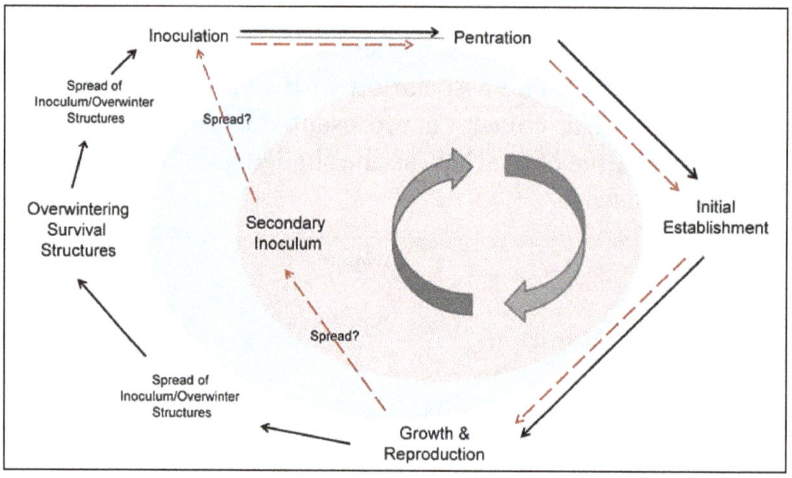

The monocyclic pathogen follows the black arrows to complete its cycle. Polycyclic pathogens follow the red arrows for the majority of the season and the black arrows at the end of the season.

In order for a disease to develop, a pathogen must be present and successfully invade plant host tissues and cells. The chain of events involved in disease development includes inoculation, penetration, infection, incubation, reproduction, and survival.

Inoculation

This describes the introduction of the plant pathogen to the host. Different pathogen groups employ different inoculation methods and are equipped with various specialized mechanisms that aid in the inoculation process. For example, some fungal pathogens release spores into the air and the spores are then spread with the aid of air currents.

Penetration

Wound sites and natural plant openings, such as stomata and hydathodes, facilitate the entrance of some plant pathogens; others have evolved unique mechanisms for direct penetration. Fungi and nematodes are able to actively penetrate host tissues and cells if environmental conditions, such as moisture and temperature, are favorable for the penetration process.

Infection

This occurs when the pathogen invades the plant tissue and establishes a parasitic relationship between itself and the plant. Viruses, bacteria, and phytoplasmas are not able to actively penetrate or enter plant host tissues. Therefore they must rely on other methods to infect plant tissues and cells. Associations with insect vectors have been established by these pathogens to aid inoculation and dispersal.

Incubation

Once inside the plant, pathogens may undergo an incubation period and remain latent for a period of time before initiating disease.

Reproduction

Plant pathogens can reproduce sexually and asexually. It is dependent on the pathogen.

Survival

Plant pathogens have evolved so they can survive prolonged periods of unfavorable weather conditions. For example, brown spot is fungal pathogens that produce spores that are dark in coloration which reduces the amount of UV light penetrating and preventing cell death. In addition, Soybean cyst nematode laid their eggs within a cuticle casing. The cuticle casing is very hard and prevents other microbes and chemicals to penetrate killing the eggs prior to hatching.

If any step is disturbed in the cycle, the disease will be less severe or fail to develop. Knowing and understanding the disease cycle for a particular disease is very helpful in managing the disease. There are two types of disease cycles, monocyclic and polycyclic.

PLANT PATHOLOGY

Plant pathology is the study of plant diseases, pathogens, and the environmental conditions that affect the plant's overall growth. The study encompasses a wide array of categories such as plant viruses, fungi, bacteria, nematodes, protozoa, viruses, virus-like organisms, parasitic plants, viroid, and oomycetes. It is also the study of pathogens, genetics, plant disease cycles, disease etiology, the economic impact of plants, diseases cycles, and the coalition of plants with humans and animals.

Plant pathology is the basic understanding of how plants grow, their lifecycle, and how they die. Through the perceptions of plant pathology, improvements on plant production, health, growth, disease resistance, and harvest can all be vastly improved. It takes into consideration many factors of a plant's life, such as pests, climate, and nutrition, to gain a greater knowledge of how to improve or combat things that may adversely take a toll on a plant's life cycle.

In some situations, plant pathology is used to improve a plant's life and increase harvest, but in other areas plant pathology is used to combat invasive plants and prevent an unbalance in the natural ecosystem that often occurs when a non-native weed invades.

PLANT PATHOGEN

A plant pathogen is a broad term that refers to any of the organisms, such as fungi, bacteria, protists, nematodes, and viruses that cause plant diseases. Plant pathogens are of interest for a number of reasons, ranging from concerns about fragile ecosystems to the desire to protect the food supply. Plant pathogens that cause plant diseases reduce a grower's ability to produce crops and can infect almost all types of plants.

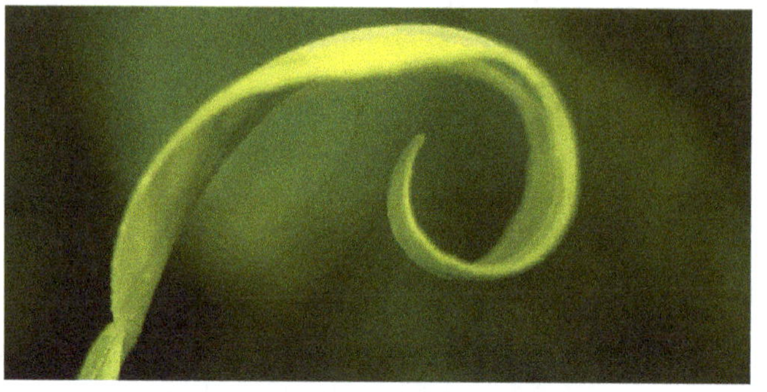

Plant Pathogen Symptoms

Plants are attacked by different groups of pathogens individually or sometimes by more than one pathogen-producing complex and more severe disease. The type of external symptoms can, in most cases, indicate the nature of the pathogen responsible for the disease. Plant pathogens can attack in a number of different ways. Some colonize the tissue in the plant, others settle on the surface of the plant, and others may go for specific areas such as the roots, stems, and leaves. Pathogens commonly cause problems like tissue death, browning, a decrease in fruiting, problems with setting flowers, and so forth. In extreme cases, they can kill the host plant.

Fungal disease symptoms	Bacterial disease symptoms	Viral disease symptoms
Birds-eye spot on berries (anthracnose) Damping off of seedlings (phytophthora) Leaf spot (septoria brown spot) Chlorosis (yellowing of leaves)	Leaf spot with yellow halo Fruit spot Canker Crown gall Sheperd's crook stem ends on woody plants	Mosaic leafpattern Crinkled leaves Yellowed leaves Plant stunting

Bioassay is a useful procedure for detection and identification of viruses and uses indicator plants which react to infection by showing characteristic symptoms. Over the last few decades, laboratory based virus test methods, such as Enzyme-Linked-Immuno-Sorbent Assay (ELISA) and Polymerase Chain Reaction (PCR), have been developed and are now being used routinely in many laboratories. Automation of these procedures enables a high throughput of samples and provides rapid results.

Plant Disease Diagnosis

Fungi, bacteria, virus, and phytoplasmas cause distinct types of symptoms in most host-pathogen interactions. However, there are some diseases which show similarity in symptoms, though they are induced by different groups of pathogens. It is often impossible to diagnose plant virus infections merely by observing host symptoms. This is not only because several viruses may induce similar symptoms in the same plant, but virus-like symptoms may also develop for physiological reasons. In addition, symptoms may be very slight and inconclusive, or plants may be infected symptomlessly.

2

Biotic Plant Pathogens

The biotic plant pathogens include various eukaryotic and prokaryotic organisms. Some of the common biotic plant pathogens are fungi, bacteria, virus, nematodes, mycoplasma, parasitic plants, magnaporthe grisea, virtoids, phytoplasma and oomycetes. This chapter discusses in detail these biotic plant pathogens.

FUNGI

Fungi and fungal like organisms (FLOs) are eukaryotic organisms that lack chlorophyll and thus do not have the ability to photosynthesize their own food. They obtain nutrients by absorption through tiny thread-like filaments called hyphae that branch in all directions throughout a substrate. A collection of hyphae is referred to as mycelium (pl. mycelia). Mycelia are the key diagnostic sign associated with diseases caused by fungi and FLOs. Most of us have seen mycelium growing on old bread or rotten fruit or vegetables and may have referred to these organisms collectively as molds or mildew.

Fungi and FLOs (indeed all pathogens) can be grouped into the following four categories based on their preference for surviving on dead or decaying organic matter versus living tissue:

1. Obligate saprophytes—Always a saprophyte. These organisms can only survive or are obliged to gain nourishment by colonizing dead or decaying organic matter. They are not parasites.
2. Obligate parasites—Always a parasite. Can only grow as a parasite on or in a living host. They cannot survive as saprophytes or be cultured in the laboratory. This is a very interesting group of pathogens in that they have a vested interest in prolonging the life of their host to increase their own viability. All viruses, downy mildews, powdery mildews, rusts and smuts are obligate parasites.
3. Facultative parasites—Usually survive as a saprophyte but have the ability to parasitize and cause disease under certain conditions. Examples include Pythium species and many bacterial pathogens.
4. Facultative saprophytes—Usually survive as a parasite but have the ability to live on dead and decaying organic matter under the right conditions. Examples include Phytophthora and Botrytis species.

Some fungi and FLOs are able to live on only one host species, while others develop on many different kinds.

Fruit rot caused by the pathogen Rhizoctonia, which also can cause damping off, root rot, and stem cankers.

Cedar-apple rust symptoms on the top and bottom of an apple leaf.

Beneficial Fungi

Fungi and FLOs can be beneficial as well as pathogenic. Beneficial fungi participate in biological cycles such as decaying dead animal and plant materials converting them into nutrients that are absorbed by living plants. Some beneficial fungi grow in a symbiotic relationship with the root cells of higher green plants; this life style is termed mycorrhizal. Roots of most cultivated plants—corn, soybeans, cotton, tobacco, peas, red clover, apples, citrus, pines, aspens, birches, turfgrass species and others—have mycorrhizal relationships with soil fungi. The mycorrhizae appear to be highly beneficial, often necessary, for optimum growth. Some beneficial fungi, such as those belonging to the genus Trichoderma, are effective biocontrol agents of plant pathogenic fungi while others, like Arthrobotrys dactyloides, have been shown to trap and parasitize plant pathogenic nematodes.

Certain fungi produce useful antibiotics and enzymes. Penicillium fungi produce the famous penicillin G, which has prevented countless deaths from bacterial infection, acting by inhibiting formation of the bacteria's cell wall. Many food-producing processes, such as the making of bread, wine, beer and cheese, are based on the activities of fungi. Most notably, mushrooms, which are fungi, are an important food for humans, animals and insects.

Pathogen Biology

Similar to all other groups of plant pathogens, fungal pathogens have developed ways to survive periods of unfavorable environmental conditions or in the absence of a susceptible host, spread, infect, grow and reproduce on and within plants. One important difference between fungi and FLOs vs. bacteria and viruses is fungi and FLOs can penetrate a host via a wound or natural opening, but they can also actively penetrate via the production of specialized hyphal structures called appresoria (sing. appresorium). Appresoria are swollen tips of hyphae that allow the fungus, through mechanical and enzymatic activity, to directly penetrate plant tissues.

Phytophthora root and stem rot of soybean.

Brown rot of a peach fruit.

Control

There are generally more options available to professional plant production specialists and growers to manage fungal and FLO diseases as compared to viral and bacterial diseases. One of the most satisfactory methods of dealing with fungal diseases is strict sanitation to eliminate the pathogenic organism, starting with the initial stages of propagation and growth of the potential host plants. Integrated management strategies for fungal and FLO diseases include the following:

Genetic Host Resistance

- Using genetically resistant species, cultivars, varieties and hybrids. In many of the major crops, cultivars resistant to prevailing diseases are available, and more are continually being developed by plant breeders. As has been discussed with other types of diseases, the use of genetically resistant plants, if available, should be the first line of defense for diseases caused by fungi and FLOs. Notable examples include: Certain hybrid potato cultivars are resistant to late blight (Phytophthora infestans); Soybean cultivars resistant to downy mildew (Peronospora manshurica) have been developed; In the United States, apple cultivars are available from the Indiana and the New York agricultural experiment stations that show high resistance or immunity to apple scab (Venturia inaequalis); In the cereal crops (oats, wheat, rye, barley), powdery mildew (Erysiphe graminis) can be controlled only by the use of resistant cultivars developed by plant breeders; Tomatoes can be grown in Fusarium-infested soils only if Fusarium-resistant cultivars are planted; Plant breeders are continuously developing wheat cultivars resistant to stem rust (Puccinia graminis tritici), but the fungus rapidly mutates, attacking the formerly resistant cultivars, requiring new resistant cultivars to be developed.

Cultural Practices

- Planting only disease-free certified seed.
- Maintaining a balanced fertility program that avoids excessive or inadequate levels of key plant nutrients.
- Maintaining an effective water management program—maintain adequate soil drainage, monitor irrigation practices, and adjust accordingly, etc.
- Ensuring proper lighting—both quality and quantity to optimize plant health—especially important in turfgrass, floricultural and ornamental nursery production systems.
- Removing crop residues by burning or burying (plowing).
- Implementing crop rotation strategies to reduce or eliminate the interaction of susceptible plants with pathogens.
- Growing crops in climates unsuitable for pathogenic fungi and FLOs.
- Careful handling of the crop (vegetables and fruits) to prevent cuts, bruises and wounding during harvest, transit and storage.
- Storage of crop products at the proper temperatures.
- Soil pasteurization (moist heat at 82 °C [180 °F] for 30 minutes).

Chemical Applications

- The use of preplant soil fumigants, the use of fungicide drenches or seed treatments with fungicides.
- Fungicide applications.
- Postharvest treatment of fruits and vegetables with fungicides.

Although the use of resistant cultivars and eradication of the pathogen through the use of cultural practices are the most satisfactory ways of dealing with diseases caused by fungi and FLOs, in many instances these measures are not possible. Often the disease appears and its development must be slowed or stopped by whatever means are available. Fungicide applications are often essential where there is a demand for plant health during environmental periods that favor pathogen growth. They are typically more effective when applied prior to the onset of disease symptoms (referred to as preventive or preventative applications). Some fungicides are effective when applied after the onset of symptoms and are said to have curative activity. In either case, fungicides must be delivered to the area of the plant where the pathogen is active to be effective. There are many different types and chemical classes of fungicides currently available. Numerous online extension-outreach and agrichemical company resources exist that provide specific fungicide recommendations for nearly every major cropping system and pathogen. Always read and follow label recommendations when applying pesticides.

Biological Control

The use of biological control organisms to suppress the activity of deleterious fungi and FLOs.

Government Regulatory Measures

The implementation of strict quarantines that exclude or restrict the introduction or movement of fungal and FLO pathogens or infected plant material.

BACTERIA

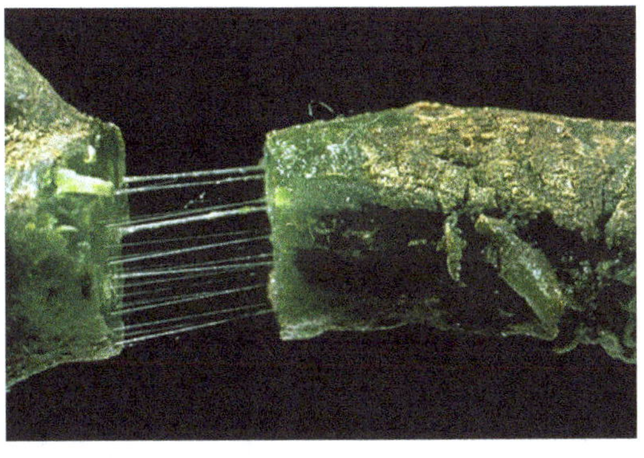

Bacterial strand test on cut stems, with bacterial slime streaming from xylem tissues.

Bacterial leaf blight on wheat.

Bacteria are microscopic, single-celled prokaryotic organisms, without a defined nucleus, that reproduce asexually by binary fission (one cell splitting into two). They occur singly or in colonies of cells. Bacteria are classified into two main groups based on cell wall structure, which can be determined by a simple staining procedure called the Gram stain. Gram negative bacteria stain red or pink and Gram positive bacteria stain purple. The difference in color is directly related to the chemical composition and structure of their cell walls. The cells can be rod-shaped, spherical, spiral-shaped or filamentous. Only a few of the latter are known to cause diseases in plants. Most bacteria are motile and have whip-like flagella that propel them through films of water.

Phytoplasmas and spiroplasmas are bacteria that lack rigid cell walls, and infect plants. Phytoplasmas are round or ovoid. As with viruses, many diseases caused by fastidious bacteria are named after the most important host plant or the one where the disease was first characterized, but some can also infect many other plants. For example, the aster yellows phytoplasma also affects other ornamentals, such as gladiolus and phlox or tomato, spinach, onion, lettuce, celery, carrots and strawberry, and many weeds.

Pathogen Biology

Crown gall, caused by Agrobacterium tumefaciens, on a burning bush.

The taxonomy of plant pathogenic bacteria is currently in flux based on recent advances on how bacteria are classified. Most plant pathogenic bacteria belong to the following genera: Erwinia, Pectobacterium, Pantoea, Agrobacterium, Pseudomonas, Ralstonia, Burkholderia, Acidovorax, Xanthomonas, Clavibacter, Streptomyces, Xylella, Spiroplasma, and Phytoplasma. Plant pathogenic bacteria cause many different kinds of symptoms that include galls and overgrowths, wilts, leaf spots, specks and blights, soft rots, as well as scabs and cankers. In contrast to viruses, which are inside host cells, walled bacteria grow in the spaces between cells and do not invade them. The means by which plant pathogenic bacteria cause disease is as varied as the types of symptoms they cause. Some plant pathogenic bacteria produce toxins or inject special proteins that lead to host cell death or they produce enzymes that break down key structural components of plant cells and their walls. An example is the production of enzymes by soft-rotting bacteria that degrade the pectin layer that holds plant cells together. Still others colonize the water-conducting xylem vessels causing the plants to wilt and die. Agrobacterium species even have the ability to genetically modify or transform their hosts and bring about the formation of cancer-like overgrowths called crown gall.

Bacteria that cause plant diseases are spread in many ways—they can be splashed about by rain or carried by the wind, birds or insects. People can unwittingly spread bacterial diseases by, for instance, pruning infected orchard trees during the rainy season. Water facilitates the entrance of bacteria carried on pruning tools into the pruning cuts. Propagation with bacteria-infected plant material is a major way pathogenic bacteria are moved over great distances. No matter how the bacterial pathogens are disseminated, they require a wound or natural opening, such as stomata, to get inside a plant host. Once inside they then kill host cells, by the means described above, so that they can grow. Between hosts they may grow harmlessly on plant surfaces and then can over-winter or survive unfavorable environmental periods or the absence of a susceptible host by either going dormant in infected tissue, infested soil or water, or in an insect vector.

Control

Bacterial diseases in plants are difficult to control. Emphasis is on preventing the spread of the bacteria rather than on curing the plant. Integrated management measures for bacterial plant pathogens include:

Genetic Host Resistance

Resistant varieties, cultivars or hybrids is the most important control procedure.

Cultural Practices

- Bacteria-free seed or propagation materials.
- Sanitation, particularly disinfestation of pruning tools.
- Crop rotation to reduce over-wintering.
- Preventing surface wounds that permit the entrance of bacteria into the inner tissues.
- Propagating only bacteria-free nursery stock.
- Prolonged exposure to dry air, heat, and sunlight will sometimes kill bacteria in plant material.

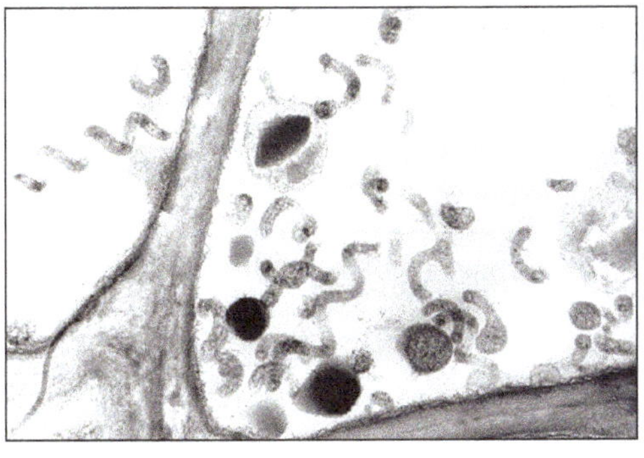

Spiroplasmas in the phloem of an infected corn plant.

Citrus canker symptoms on fruit.

Chemical Applications

- Applications of copper-containing compounds or Bordeaux mixture (copper sulfate and lime).
- Antibiotics: Streptomycin and/or oxytetracycline may also help kill or suppress plant pathogenic bacteria prior to infection and reduce spread of the disease, but they will not cure plants that are already diseased.

- Antibiotics are also used to treat diseases caused by fastidious vascular bacteria. Phytoplasmas and spiroplasmas are susceptible to certain antibiotics, particularly tetracycline, which has been used to treat pear trees with the pear decline disease. Tetracycline must be injected into mature trees on a routine or therapeutic schedule to be effective and even then only appears to suppress the development of symptoms rather than curing the infected plant. Applications made during the early stages of infection tend to be more effective than in the later stages of disease development.
- Insect control will help to eliminate vectors or reduce feeding wounds that can provide points of entry.

Biological Control

The use of antagonistic or biological control products may also be effective for managing bacterial diseases of plants.

VIRUS

Plant viruses are viruses that affect plants. Like all other viruses, plant viruses are obligate intracellular parasites that do not have the molecular machinery to replicate without a host. Plant viruses can be pathogenic to higher plants.

Most plant viruses are rod-shaped, with protein discs forming a tube surrounding the viral genome; isometric particles are another common structure. They rarely have an envelope. The great majority have an RNA genome, which is usually small and single stranded (ss), but some viruses have double-stranded (ds) RNA, ssDNA or dsDNA genomes. Although plant viruses are not as well understood as their animal counterparts, one plant virus has become iconic: *tobacco mosaic virus* (TMV), the first virus to be discovered. This and other viruses cause an estimated U.S$60 billion loss in crop yields worldwide each year. Plant viruses are grouped into 73 genera and 49 families. However, these figures relate only to cultivated plants, which represent only a tiny fraction of the total number of plant species. Viruses in wild plants have been relatively little studied, but the interactions between wild plants and their viruses often do not appear to cause disease in the host plants.

To transmit from one plant to another and from one plant cell to another, plant viruses must use strategies that are usually different from animal viruses. Plants do not move, and so plant-to-plant transmission usually involves vectors (such as insects). Plant cells are surrounded by solid cell walls, therefore transport through plasmodesmata is the preferred path for virions to move between plant cells. Plants have specialized mechanisms for transporting mRNAs through plasmodesmata, and these mechanisms are thought to be used by RNA viruses to spread from one cell to another. Plant defenses against viral infection include, among other measures, the use of siRNA in response to dsRNA. Most plant viruses encode a protein to suppress this response. Plants also reduce transport through plasmodesmata in response to injury.

The discovery of plant viruses causing disease is often accredited to A. Mayer (1886) working in the Netherlands demonstrated that the sap of mosaic obtained from tobacco leaves developed mosaic symptom when injected in healthy plants. However the infection of the sap was destroyed when it was boiled. He thought that the causal agent was the bacteria. However, after larger inoculation with a large number of bacteria, he failed to develop a mosaic symptom.

In 1898, Martinus Beijerinck, who was a Professor of Microbiology at the Technical University the Netherlands, put forth his concepts that viruses were small and determined that the "mosaic disease" remained infectious when passed through a Chamberland filter-candle .This was in contrast to bacteria microorganisms, which were retained by the filter. Beijerinck referred to the infectious filtrate as a "contagium vivum fluidum", thus the coinage of the modern term "virus".

After the initial discovery of the 'viral concept' there was need to classify any other known viral diseases based on the mode of transmission even though microscopic observation proved fruitless. In 1939 Holmes published a classification list of 129 plant viruses. This was expanded and in 1999 there were 977 officially recognized, and some provisional, plant virus species.

The purification (crystallization) of TMV was first performed by Wendell Stanley, who published his findings in 1935, although he did not determine that the RNA was the infectious material. However, he received the Nobel Prize in Chemistry in 1946. In the 1950s a discovery by two labs simultaneously proved that the purified RNA of the TMV was infectious which reinforced the argument. The RNA carries genetic information to code for the production of new infectious particles.

More recently virus research has been focused on understanding the genetics and molecular biology of plant virus genomes, with a particular interest in determining how the virus can replicate, move and infect plants. Understanding the virus genetics and protein functions has been used to explore the potential for commercial use by biotechnology companies. In particular, viral-derived sequences have been used to provide an understanding of novel forms of resistance. The recent boom in technology allowing humans to manipulate plant viruses may provide new strategies for production of value-added proteins in plants.

Structure

Viruses are extremely small and can only be observed under an electron microscope. The structure of a virus is given by its coat of proteins, which surround the viral genome. Assembly of viral particles takes place spontaneously.

Over 50% of known plant viruses are rod-shaped (flexuous or rigid). The length of the particle is normally dependent on the genome but it is usually between 300–500 nm with a diameter of 15–20 nm. Protein subunits can be placed around the circumference of a circle to form a disc. In the presence of the viral genome, the discs are stacked, then a tube is created with room for the nucleic acid genome in the middle.

The second most common structure amongst plant viruses are isometric particles. They are 25–50 nm in diameter. In cases when there is only a single coat protein, the basic structure consists of 60 T subunits, where T is an integer. Some viruses may have 2 coat proteins that associate to form an icosahedral shaped particle.

There are three genera of *Geminiviridae* that consist of particles that are like two isometric particles stuck together.

A very small number of plant viruses have, in addition to their coat proteins, a lipid envelope. This is derived from the plant cell membrane as the virus particle buds off from the cell.

Transmission of Plant Viruses

Through Sap

Viruses can be spread by direct transfer of sap by contact of a wounded plant with a healthy one. Such contact may occur during agricultural practices, as by damage caused by tools or hands, or naturally, as by an animal feeding on the plant. Generally TMV, potato viruses and cucumber mosaic viruses are transmitted via sap.

Insects

Plant viruses need to be transmitted by a vector, most often insects such as leafhoppers. One class of viruses, the Rhabdoviridae, has been proposed to actually be insect viruses that have evolved to replicate in plants. The chosen insect vector of a plant virus will often be the determining factor in that virus's host range: it can only infect plants that the insect vector feeds upon. This was shown in part when the old world white fly made it to the United States, where it transferred many plant viruses into new hosts. Depending on the way they are transmitted, plant viruses are classified as non-persistent, semi-persistent and persistent. In non-persistent transmission, viruses become attached to the distal tip of the stylet of the insect and on the next plant it feeds on, it inoculates it with the virus. Semi-persistent viral transmission involves the virus entering the foregut of the insect. Those viruses that manage to pass through the gut into the haemolymph and then to the salivary glands are known as persistent. There are two sub-classes of persistent viruses: propagative and circulative. Propagative viruses are able to replicate in both the plant and the insect (and may have originally been insect viruses), whereas circulative can not. Circulative viruses are protected inside aphids by the chaperone protein symbionin, produced by bacterial symbionts. Many plant viruses encode within their genome polypeptides with domains essential for transmission by insects. In non-persistent and semi-persistent viruses, these domains are in the coat protein and another protein known as the helper component. A bridging hypothesis has been proposed to explain how these proteins aid in insect-mediated viral transmission. The helper component will bind to the specific domain of the coat protein, and then the insect mouthparts — creating a bridge. In persistent propagative viruses, such as tomato spotted wilt virus (TSWV), there is often a lipid coat surrounding the proteins that is not seen in other classes of plant viruses. In the case of TSWV, 2 viral proteins are expressed in this lipid envelope. It has been proposed that the viruses bind via these proteins and are then taken into the insect cell by receptor-mediated endocytosis.

Plasmodiophorids

A number of virus genera are transmitted, both persistently and non-persistently, by soil borne zoosporic protozoa. These protozoa are not phytopathogenic themselves, but parasitic. Transmission of the virus takes place when they become associated with the plant roots. Examples include *Polymyxa graminis*, which has been shown to transmit plant viral diseases in cereal crops and *Polymyxa betae* which transmits Beet necrotic yellow vein virus. Plasmodiophorids also create wounds in the plant's root through which other viruses can enter.

Seed and Pollen Borne Viruses

Plant virus transmission from generation to generation occurs in about 20% of plant viruses. When viruses are transmitted by seeds, the seed is infected in the generative cells and the virus is maintained in the germ cells and sometimes, but less often, in the seed coat. When the growth and development of plants is delayed because of situations like unfavorable weather, there is an increase in the amount of virus infections in seeds. There does not seem to be a correlation between the location of the seed on the plant and its chances of being infected. Little is known about the mechanisms involved in the transmission of plant viruses via seeds, although it is known that it is environmentally influenced and that seed transmission occurs because of a direct invasion of the embryo via the ovule or by an indirect route with an attack on the embryo mediated by infected gametes. These processes can occur concurrently or separately depending on the host plant. It is unknown how the virus is able to directly invade and cross the embryo and boundary between the parental and progeny generations in the ovule. Many plants species can be infected through seeds including but not limited to the families Leguminosae, Solanaceae, Compositae, Rosaceae, Cucurbitaceae, Gramineae. Bean common mosaic virus is transmitted through seeds.

Direct Plant-to-Human Transmission

Researchers from the University of the Mediterranean in Marseille, France have found tenuous evidence that suggest a virus common to peppers, the Pepper Mild Mottle Virus (PMMoV) may have moved on to infect humans. This is a very rare and highly unlikely event as, to enter a cell and replicate, a virus must "bind to a receptor on its surface, and a plant virus would be highly unlikely to recognize a receptor on a human cell. One possibility is that the virus does not infect human cells directly. Instead, the naked viral RNA may alter the function of the cells through a mechanism similar to RNA interference, in which the presence of certain RNA sequences can turn genes on and off," according to Virologist Robert Garry from the Tulane University in New Orleans, Louisiana.

Translation of Plant Viral Proteins

75% of plant viruses have genomes that consist of single stranded RNA (ssRNA). 65% of plant viruses have +ssRNA, meaning that they are in the same sense orientation as messenger RNA but 10% have -ssRNA, meaning they must be converted to +ssRNA before they can be translated. 5% are double stranded RNA and so can be immediately translated as +ssRNA viruses. 3% require a reverse transcriptase enzyme to convert between RNA and DNA. 17% of plant viruses are ssDNA and very few are dsDNA, in contrast a quarter of animal viruses are dsDNA and three quarters of bacteriophage are dsDNA. Viruses use the plant ribosomes to produce the 4-10 proteins encoded by their genome. However, since many of the proteins are encoded on a single strand (that is, they are polycistronic) this will mean that the ribosome will either only produce one protein, as it will terminate translation at the first stop codon, or that a polyprotein will be produced. Plant viruses have had to evolve special techniques to allow the production of viral proteins by plant cells.

5′ Cap

For translation to occur, eukaryotic mRNAs require a 5′ Cap structure. This means that viruses must also have one. This normally consists of 7MeGpppN where N is normally adenine or guanine. The viruses encode a protein, normally a replicase, with a methyltransferase activity to allow this.

Some viruses are cap-snatchers. During this process, a ^{7m}G-capped host mRNA is recruited by the viral transcriptase complex and subsequently cleaved by a virally encoded endonuclease. The resulting capped leader RNA is used to prime transcription on the viral genome.

However some plant viruses do not use cap, yet translate efficiently due to cap-independent translation enhancers present in 5′ and 3′ untranslated regions of viral mRNA.

Some viruses (e.g. tobacco mosaic virus (TMV)) have RNA sequences that contain a "leaky" stop codon. In TMV 95% of the time the host ribosome will terminate the synthesis of the polypeptide at this codon but the rest of the time it continues past it. This means that 5% of the proteins produced are larger than and different from the others normally produced, which is a form of translational regulation. In TMV, this extra sequence of polypeptide is an RNA polymerase that replicates its genome.

Production of Sub-Genomic RNAs

Some viruses use the production of subgenomic RNAs to ensure the translation of all proteins within their genomes. In this process the first protein encoded on the genome, and this the first to be translated, is a replicase. This protein will act on the rest of the genome producing negative strand sub-genomic RNAs then act upon these to form positive strand sub-genomic RNAs that are essentially mRNAs ready for translation.

Segmented Genomes

Some viral families, such as the *Bromoviridae* instead opt to have multipartite genomes, genomes split between multiple viral particles. For infection to occur, the plant must be infected with all particles across the genome. For instance *Brome mosaic virus* has a genome split between 3 viral particles, and all 3 particles with the different RNAs are required for infection to take place.

Polyprotein Processing

This strategy is adopted by viral genera such as the Potyviridae and Tymoviridae. The ribosome translates a single protein from the viral genome. Within the polyprotein is an enzyme (or enzymes) with proteinase function that is able to cleave the polyprotein into the various single proteins or just cleave away the protease, which can then cleave other polypeptides producing the mature proteins.

NEMATODES

Nematodes are simple, multi-cellular animals—typically containing 1,000 cells or less. They are worm-like in appearance, but are taxonomically distinct from earthworms, wireworms or flatworms. They are bilaterally symmetrical, soft-bodied (no skeleton), non-segmented round worms. Most nematode species that attack plants are microscopic. The basic body plan of a nematode is a "tube within a tube." Nematodes feed on other microorganisms and plants like bacteriovores, fungivores, omnivores, predators, and plant parasites.

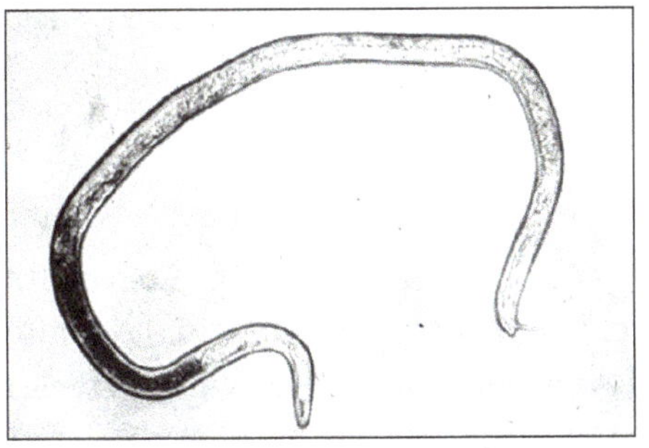

Adult root-knot nematode.

Plant parasitic nematodes may attack the roots, stem, foliage and flowers of plants. All plant parasitic nematodes have piercing mouthparts called stylets. The presence of a stylet is the key diagnostic sign differentiating plant parasitic nematodes from all other types of nematodes. The bacterial-feeding nematode, Caenorhabditis elegans, is one of the best-understood animals on earth. It was the first animal to have its entire genome completely sequenced. The study of C. elegans has led to many new insights into animal development, neurobiology and behavior.

Signs and Symptoms

Typical root symptoms indicating nematode attack are root knots or galls, root lesions, excessive root branching, injured root tips and stunted root systems. Symptoms on the above-ground plant parts indicating root infection are a slow decline of the entire plant, wilting even with ample soil moisture, foliage yellowing and fewer and smaller leaves. These are, in fact, the symptoms that would appear in plants deprived of a properly functioning root system. Bulb and stem nematodes produce stem swellings and shortened internodes. Bud and leaf nematodes distort and kill bud and leaf tissue. In some cases, such as with SCN, yield loss may take place with no visible symptoms.

Dissemination

Parasitic nematodes are readily spread by any physical means that can move soil particles about—equipment, tools, shoes, birds, insects, dust, wind and water. In addition, the movement of nematode-infested plants or plant parts will spread the parasites.

Control

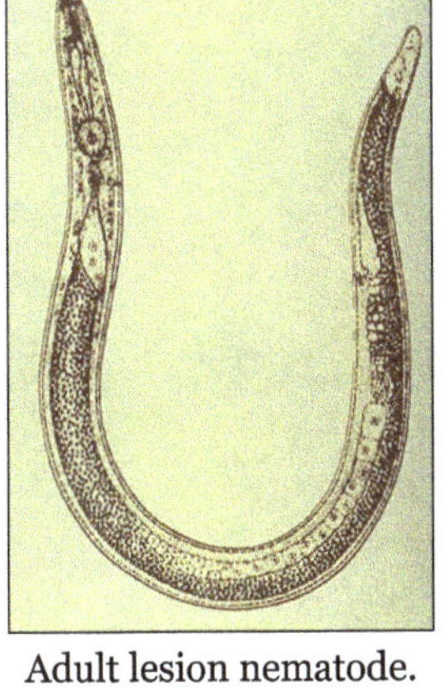

Adult lesion nematode.

Various methods are available to reduce crop losses from nematodes:

Genetic Host Resistance

Plant resistant species and cultivars. For example, in an area with soil heavily infested with the root-knot nematode, plant apricots, cherries, apples, pears or plums, which are resistant, rather than peaches or nectarines, which are highly susceptible. (A root-knot nematode-resistant peach rootstock called 'Nemaguard' developed by USDA plant breeders is available, thus permitting peach production even on infested soils.) Certain vegetable crops—sweet corn, asparagus, and cabbage—are resistant to root-knot nematodes whereas radishes are susceptible. Resistant ornamentals include the African marigold, azalea, camellia and oleander. In Long Island, New York, where the golden nematode is a serious problem for potato production, resistant cultivars are available. Similarly, soybean varieties resistant to soybean cyst nematode (Heterodera glycines) are also available.

Cultural Practices

- Use only nematode-free nursery stock for planting. In most countries, government nursery inspectors will condemn and destroy any nursery stock showing evidence of nematode infestation.

- In nursery operations, use benches raised off the ground and pot plants only into pasteurized soil mixes. Keep containers, bins, benches and flats clean. Fumigate outdoor growing fields where nursery stock will be grown.

- Rotate crops to control certain nematodes. Rotation is useful for types that have a narrow host range, such as sugar beets attacked by the sugar beet cyst nematode. Where the crop value is too low to justify large-scale soil fumigation, crop rotation is the only practical method of nematode control.

- Use cover crops that reduce nematode damage. Cover crops can improve soil structure and fertility, decrease soil erosion, be used as animal feed, and suppress weeds, insects and pathogens. Examples of cover crops that have been shown to suppress nematodes include cowpea, rapeseed, velvet bean and sudangrass.

Chemical Applications

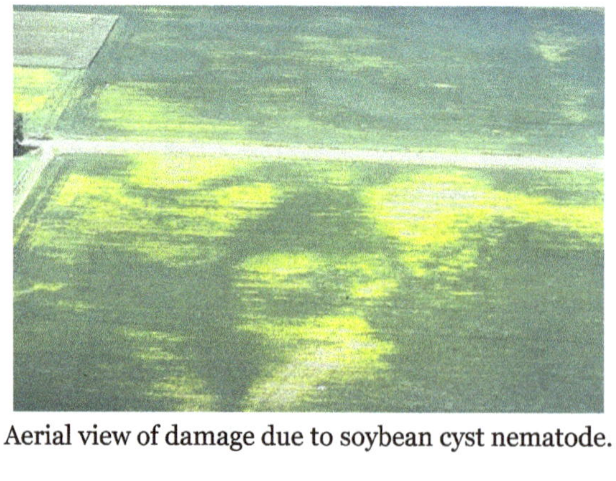

Aerial view of damage due to soybean cyst nematode.

- Treat the soil area with a fumigant before planting. Soil mixes for container-grown plants can either be treated with a fumigant or steam-pasteurized at 82 °C (180 °F) for about 30 minutes. This method is too expensive for field crops other than commercial strawberry fields. The impending loss of methyl bromide may seriously affect the crops where it is used.
- Use nematicides in certain cases. All nematicides are poisonous and must be used carefully, following the directions on the containers exactly. Most such materials will injure or kill plants if applied too close to their root zones. As the number of commercially available nematicides decreases, greater emphasis has been placed on the development of alternative IPM practices.

Biological Control

Although not widely available, scientists have explored the use of antagonistic fungi like Arthrobotrys dactyloides to trap and parasitize plant pathogenic nematodes. Pasteuria penetrans, a bacterial parasite, can also be used as biological control.

Government Regulatory Measures

White female soybean cyst nematodes on root surface.

Avoid importing soil (or plants with soil on their roots) from areas that could be loaded with a dangerous nematode species new to the area. United States plant importation regulations forbid the introduction of plants with soil on their roots from other countries.

MYCOPLASMA

Mycoplasmas are similar to bacteria, and are transmitted by insects (such as leafhoppers) that pierce and suck.

They are widespread in southern Europe, and cause symptoms known as phyllody, in which flowerstalks grow large and flattened as if they were leaves.

Mycoplasmas or Plant Disease Mollicutes

There are some diseases which until 1967 were attributed to viruses, although no viral particles had been found in the affected plants. Even so, given the observation and detection methods available at the time, this was the most probable hypothesis. In 1967 it was conclusively shown that they were nothing to do with viruses.

With the coming of the electron microscope it was possible to identify some polymorphic structures which were present in the phloem of the sick plants and not in that of healthy ones. These structures had certain similarities to prokaryotic micro-organisms known as mycoplasmas, which had long been recognised as agents of infection in animals or as saprophytes on mucus.

These mycoplamsas are similar to bacteria except that they have no true cell wall. They are generally assigned an intermediate place between bacteria and viruses.

Following this discovery a number of "inexplicable" plant diseases have been attributed to these organisms of the mycoplasma type, also known as Mollicutes.

They take a great many forms and are involved in many diseases with widely differing symptoms. On the other hand their form of organisation means that these mycoplasmas cannot evolve outside certain special cellular niches within the host, in this case the cells of the phloem.

Manner of Transmission

Plant disease mycoplasmas are carried from a diseased plant to a healthy one by piercing and sucking insects, often of the Homopteran order such as leafhoppers.

These insects suck up the sap made by an infected plant and with it the mycoplasmas it contains. Mycoplasma transmission by leafhopper is circulating and persistent. It requires a period of latency corresponding to its circulation and multiplication within the vector insect, which then remains infectious for the rest of its life. The mycoplasmas are not however passed on to the offspring.

No mycoplasma can be transmitted from plant to plant by physical contact, nor in seed.

Symptoms

The diseases caused are known as 'mycoplasmoses', or 'mycoplasmainfections', and show very diverse symptoms.

In most cases one or more of the following symptoms are observed:

- Yellowing;
- Various growth disorders such as dwarfism, polyphylly, witches' brooms (abnormal development of axillary buds);
- Colour disorders such as variegation;
- Flower deformation: virescence (leaf-like aspect of parts of flowers) and phyllody (elongation of the gynoecium as leaf-like structures).

These disorders are essentially the result of disturbance in the functioning of the phloem: the transport and transfer of the energy-carrying molecules and mineral salts are upset and the action of growth and development factors (plant hormones) is disturbed.

Mycoplasma.

Mycoplasma.

Mycoplasma.

Countermeasures

The symptoms are not very specific, especially as there are a number of factors which can complicate the diagnosis: particular temperature conditions may mask the symptoms, for instance; so may active phases of growth.

What is more, the epidemiology of these diseases is greatly affected by the behaviour of the vector. This may change considerably as a function of changes in climatic and horticultural conditions (temperature, timing of planting, irrigation, plant density).

Control measures consist of the use of chemicals active against the insect vectors of the mycoplasmas, in other words against the Homopterans. It is also essential to destroy the affected plants. There are antibiotics which show some effect, but their use in agriculture is prohibited.

PARASITIC PLANTS

A parasitic plant is a plant that derives some or all of its nutritional requirement from another living plant. They make up about 1% of angiosperms and are in almost every biome in the world. All parasitic plants have modified roots, called haustoria, which penetrate the host plants, connecting them to the conductive system – either the xylem, the phloem, or both. For example, plants like *Striga* or *Rhinanthus* connect only to the xylem, via xylem bridges (xylem-feeding). Alternately, plants like *Cuscuta* and *Orobanche* connect only to the phloem of the host (phloem-feeding). This provides them with the ability to extract water and nutrients from the host. Parasitic plants are classified depending on where the parasitic plant latches onto the host and the amount of nutrients it requires. Some parasitic plants are able to locate their host plants by detecting chemicals in the air or soil given off by host shoots or roots, respectively. About 4,500 species of parasitic plant in approximately 20 families of flowering plants are known.

Classification

Parasitic plants occur in multiple plant families, indicating that the evolution is polyphyletic. Some families are comprised mostly of parasitic representatives such as Balanophoraceae, while other families have only a few representatives. One example is the North American *Monotropa uniflora* (Indian pipe or corpse plant) which is a member of the heath family, Ericaceae, better known for its members blueberries, cranberries, and rhododendrons.

Parasitic plants are characterized as follows:

- Obligate parasite – A parasite that cannot complete its life cycle without a host.
- Facultative parasite – A parasite that can complete its life cycle independent of a host.
- Stem parasite – A parasite that attaches to the host stem.
- Root parasite – A parasite that attaches to the host root.
- Hemiparasite – A plant parasitic under natural conditions, but photosynthetic to some degree. Hemiparasites may just obtain water and mineral nutrients from the host plant; many obtain at least part of their organic nutrients from the host as well.
- Holoparasite - A parasitic plant that derives all of its fixed carbon from the host plant. Commonly lacking chlorophyll, holoparasites are often colors other than green.

Mistletoe, an obligate stem hemiparasite.

For hemiparasites, one from each of the three sets of terms can be applied to the same species, e.g. given below.

- *Nuytsia floribunda* (Western Australian Christmas tree) is an obligate root hemiparasite.
- *Rhinanthus* (e.g. Yellow rattle) is a facultative root hemiparasite.
- Mistletoe is an obligate stem hemiparasite.

Holoparasites are always obligate so only two terms are needed, e.g. given below.

- Dodder is a stem holoparasite.
- *Hydnora* spp. are root holoparasites.

Plants usually considered holoparasites include broomrape, dodder, *Rafflesia*, and the Hydnoraceae. Plants usually considered hemiparasites include *Castilleja*, mistletoe, Western Australian Christmas tree, and yellow rattle.

Evolution of Parasitic Behavior

Striga witchweeds (white, centre, attached to roots of host) are economically important pests of the crop plants that they parasitise.

Parasitic behavior evolved in angiosperms roughly 12-13 times independently, a classic example of convergent evolution. Roughly 1% of all angiosperm species are parasitic, with a large degree of host dependence. The taxonomic family *Orobanchaceae* (encompassing the genera *Tryphysaria*, *Striga*, and *Orobanche*) is the only family that contains both holoparasitic and hemiparasitic species, making it a model group for studying the evolutionary rise of parasitism. The remaining groups contain only hemiparasites or holoparasites.

The evolutionary event which gave rise to parasitism in plants was the development of haustoria. The first, most ancestral, haustoria are thought to be similar to that of the facultative hemiparasites within *Tryphysaria*, lateral haustoria develop along the surface of the roots in these species. Later evolution led to the development of terminal or primary haustoria at the tip of the juvenile radicle, seen in obligate hemiparasitic species within *Striga*. Lastly, obligate holoparasitic behavior originated with the loss of the photosynthetic process, seen in the genus *Orobanche*.

To maximize resources, many parasitic plants have evolved 'self-incompatibility', to avoid parasitizing themselves. Others such as *Triphysaria* usually avoid parasitizing other members of their species, but some parasitic plants have no such limits. The albino redwood is a mutant *Sequoia sempervirens* that produces no chlorophyll; they live on sugars from neighbouring trees, usually the parent tree from which they have grown (via a somatic mutation).

Seed Germination

Parasitic plants germinate in a variety of ways. These means can either be chemical or mechanical and the means used by seeds often depends on whether or not the parasites are root parasites or stem parasites. Most parasitic plants need to germinate in close proximity to their host plants because their seeds are limited in the amount of resources necessary to survive without nutrients from their host plants. Resources are limited due in part to the fact that most parasitic plants are not able to use autotrophic nutrition to establish the early stages of seeding.

Root parasitic plant seeds tend to use chemical cues for germination. In order for germination to occur, seeds need to be fairly close to their host plant. For example, the seeds of witchweed (*Striga asiatica*) need to be within 3 to 4 millimeters (mm) of its host in order to pick up chemical signals in the soil to signal germination. This range is important because *Striga asiatica* will only grow about 4 mm after germination. Chemical compound cues sensed by parasitic plant seeds are from host plant root exudates that are leached in close proximity from the host's root system into the surrounding soil. These chemical cues are a variety of compounds that are unstable and rapidly degraded in soil and are present within a radius of a few meters of the plant exuding them. Parasitic plants germinate and follow a concentration gradient of these compounds in the soil toward the host plants if close enough. These compounds are called strigolactones. Strigolactone stimulates ethylene biosynthesis in seeds causing them to germinate.

There are a variety of chemical germination stimulants. Strigol was the first of the germination stimulants to be isolated. It was isolated from a non-host cotton plant and has been found in true host plants such as corn and millets. The stimulants are usually plant specific, examples of other germination stimulants include sorgolactone from sorghum, orobanchol and alectrol from red clover, and 5-deoxystrigol from *Lotus japonicus*. Strigolactones are apocarotenoids that are produced via the carotenoid pathway of plants. Strigolactones and mycorrhizal fungi have a relationship in which Strigolactone also cues the growth of mycorrhizal fungus.

Stem parasitic plants, unlike most root parasites, germinate using the resources inside their endosperms and are able to survive for some time. For example, the dodders (*Cuscuta* spp.) drop their seeds to the ground; these may remain dormant for up to five years before they find a host plant nearby. Using the resources in the seed endosperm, dodder is able to germinate. Once germinated, the plant has 6 days to find and establish a connection with its host plant before its resources run out. Dodder seeds germinate above ground and then the plant sends out stems in search of its host plant reaching up to 6 cm before it dies. It is believed that the plant uses two methods of finding a host. The stem detects its host plant's scent and orients itself in that direction. Scientists used volatiles from tomato plants (α-pinene, β-myrcene, and β-phellandrene) to test the reaction of *C. pentagona* and found that the stem orients itself in the direction of the odor. Some studies suggest that by using light reflecting from nearby plants dodders are able to select host with higher sugar because of the levels of chlorophyll in the leaves. Once the dodder finds its host, it wraps itself around the host plants stem. Using adventitious roots, the dodder taps into the host plant's stem with a haustorium, an absorptive organ within the host plant vascular tissue. Dodder makes several of these connections with the host as it moves up the plant.

Seed Dispersal

There are several methods of seed dispersal, but all the strategies aim to put the seed in direct contact with, or within a critical distance of, the host.

- The *Cuscuta* seedling can live for 3–7 weeks and extend out 35 cm in search of the host before it dies. This is because the *Cuscuta* seed is large and has stored nutrients to sustain its life. This is also useful for seeds that get digested by animals and excreted out.

- Mistletoe use a sticky seed for dispersal. The seed sticks to nearby animals and birds and then comes into direct contact with the host.
- *Arceuthobium* seeds have a similarly sticky seed as the mistletoe but they do not rely on animals and birds, they mainly disperse by fruit explosiveness. Once the seed makes contact with the host rain water can help position the seed into a suitable position.
- Some seeds detect and respond to chemical stimulations produced in the host's roots and start to grow towards the host.

Obstacles of Attaching to a Host

A parasitic plant has many obstacles to overcome in order to attach to the host. Distance from the host and stored nutrients are only some of the problems, the host's defenses are an obstacle to overcome itself. The first hurdle is penetrating the host, the host has systems to reinforce the cell wall by protein cross-linking so that it stops the parasitic progress at the cortex of the host's roots. The second hurdle is the host's ability to secrete germination inhibitors. This prevents germination of the parasitic seed. The third hurdle is the host's ability to create a toxic environment for where the parasitic plant attaches to. The host secretes phenolic compounds into the apoplast the creates a toxic environment for the parasitic plant eventually killing it. The fourth hurdle is the host's ability to ruin the tubercle using gums and gels or injecting toxins into the tubercle.

Host Range

Some parasitic plants are generalists and parasitize many different species, even several different species at once. Dodder (*Cassytha* spp. *Cuscuta* spp.) and red rattle (*Odontites vernus*) are generalist parasites. Other parasitic plants are specialists that parasitize a few or even just one species. Beech drops (*Epifagus virginiana*) is a root holoparasite only on American beech (*Fagus grandifolia*). *Rafflesia* is a holoparasite on the vine *Tetrastigma*. Plants such as *Pterospora* become parasites of mycorrhizal fungi. There is evidence that parasites also practice self-discrimination, species of *Tryphysaria* experience reduced haustorium development in the presence of other *Tryphysaria*. Although, the mechanism for self-discrimination in parasites is not yet known.

Aquatic Parasitic Plants

Parasitism also evolved within aquatic species of plants and algae. Parasitic marine plants are described as benthic, meaning that they are sedentary or attached to another structure. Plants and algae that grow on the host plant, using it as an attachment point are given the designation epiphytic (epilithic is the name given to plants/algae that use rocks or boulders for attachment), while not necessarily parasitic, some species occur in high correlation with a certain host species, suggesting that they rely on the host plant in some way or another. In contrast, endophytic plants and algae grow inside their host plant, these have a wide range of host dependence from obligate holoparasites to facultative hemiparasites.

Marine parasites occur as a higher proportion of marine flora in temperate rather than tropical waters. While no full explanation for this is available, many of the potential host plants such as kelp and other macroscopic brown algae are generally restricted to temperate areas. Roughly 75% of parasitic red algae infect hosts in the same taxonomic family as themselves, these are given the designation adelphoparasites. Other marine parasites, deemed endozoic, are parasites of marine invertebrates (molluscs, flatworms, sponges) and can be either holoparasitic or hemiparasitic, some retaining the ability to photosynthesize after infection.

Importance

Species within *Orobanchaceae* are some of the most economically destructive species on Earth. Species of *Striga* alone are estimated to cost billions of dollars a year in crop yield loss annually, infesting over 50 million hectares of cultivated land within sub-Saharan Africa alone. *Striga* can infest both grasses and grains, including corn, rice and Sorghum, undoubtedly some of the most important food crops. *Orobanche* also threatens a wide range of important crops, including peas, chickpeas, tomatoes, carrots, and varieties of the genus *Brassica* (e.g. cabbage, lettuce, and broccoli). Yield loss from *Orobanche* can reach 100% and has caused farmers in some regions of the world to abandon certain staple crops and begin importing others as an alternative. Much research has been devoted to the control of *Orobanche* and *Striga* species, which are even more devastating in developing areas of the world, though no method has been found to be entirely successful.

- Mistletoes cause economic damage to forest and ornamental trees.
- *Rafflesia arnoldii* produces the world's largest flowers at about one meter in diameter. It is a tourist attraction in its native habitat.
- Sandalwood trees (*Santalum* species) have many important cultural uses and their fragrant oils have high commercial value.
- Indian paintbrush (*Castilleja linariaefolia*) is the state flower of Wyoming.
- The oak mistletoe (*Phoradendron serotinum*) is the floral emblem of Oklahoma.
- A few other parasitic plants are occasionally cultivated for their attractive flowers, such as *Nuytsia* and broomrape.
- Parasitic plants are important in research, especially on the loss of photosynthesis during evolution.
- A few dozen parasitic plants have occasionally been used as food by people.
- Western Australian Christmas tree (*Nuytsia floribunda*) sometimes damages underground cables. It mistakes the cables for host roots and tries to parasitize them using its sclerenchymatic guillotine.

Some parasitic plants are destructive while some have positive influences in their communities. Some parasitic plants damage invasive species more than native species. This results in the reduced damage of invasive species in the community.

Newly emergent snow plant (*Sarcodes sanguinea*), a flowering plant parasitic on mycorrhizal fungi.

In many regions, including the Nepal Eastern Himalayas, parasitic plants are used for medicinal and ritual purposes.

Plants Parasitic on Fungi

About 400 species of flowering plants, plus one gymnosperm (*Parasitaxus usta*), are parasitic on mycorrhizal fungi. This effectively gives these plants the ability to become associated with many of the other plants around them. They are termed myco-heterotrophs. Some myco-heterotrophs are Indian pipe (*Monotropa uniflora*), snow plant (*Sarcodes sanguinea*), underground orchid (*Rhizanthella gardneri*), bird's nest orchid (*Neottia nidus-avis*), and sugarstick (*Allotropa virgata*). Within the taxonomic family *Ericaceae*, known for extensive mycorrhizal relationships, there are the Monotropoids. The Monotropoids include the genera *Monotropa*, *Monotropsis*, and *Pterospora* among others. Myco-heterotrophic behavior is commonly accompanied by the loss of chlorophyll.

MAGNAPORTHE GRISEA

Magnaporthe grisea, also known as rice blast fungus, rice rotten neck, rice seedling blight, blast of rice, oval leaf spot of graminea, pitting disease, ryegrass blast, and Johnson spot, is a plant-pathogenic fungus that causes a serious disease affecting rice. It is now known that *M. grisea* consists of a cryptic species complex containing at least two biological species that have clear genetic differences and do not interbreed. Complex members isolated from *Digitaria* have been more narrowly defined as *M. grisea*. The remaining members of the complex isolated from rice and a variety of other hosts have been renamed *Magnaporthe oryzae*. Confusion on which of these two names to use for the rice blast pathogen remains, as both are now used by different authors.

Members of the *Magnaporthe grisea* complex can also infect other agriculturally important cereals including wheat, rye, barley, and pearl millet causing diseases called blast disease or blight disease. Rice blast causes economically significant crop losses annually. Each year it is estimated to destroy enough rice to feed more than 60 million people. The fungus is known to occur in 85 countries worldwide.

Hosts and Symptoms

M. grisea is an ascomycete fungus. It is an extremely effective plant pathogen as it can reproduce both sexually and asexually to produce specialized infectious structures known as appressoria that infect aerial tissues and hyphae that can infect root tissues.

Lesions on rice leaves caused by infection with *M. grisea*.

Rice blast lesions on plant nodes.

Rice blast has been observed on rice strains M-201, M-202, M-204, M-205, M-103, M-104, S-102, L-204, Calmochi-101, with M-201 being the most vulnerable. Initial symptoms are white to gray-green lesions or spots with darker borders produced on all parts of the shoot, while older lesions are elliptical or spindle-shaped and whitish to gray with necrotic borders. Lesions may enlarge and coalesce to kill the entire leaf. Symptoms are observed on all above-ground parts of the plant. Lesions can be seen on the leaf collar, culm, culm nodes, and panicle neck node. Internodal infection of the culm occurs in a banded pattern. Nodal infection causes the culm to break at the infected node (rotten neck). It also affects reproduction by causing the host to produce fewer seeds. This is caused by the disease preventing maturation of the actual grain.

Disease Cycle

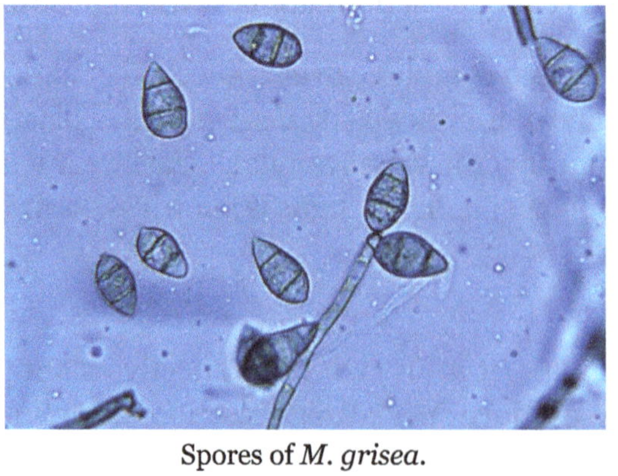

Spores of *M. grisea*.

The pathogen infects as a spore that produces lesions or spots on parts of the rice plant such as the leaf, leaf collar, panicle, culm and culm nodes. Using a structure called an appressorium, the pathogen penetrates the plant. The pathogen is able to move between the plant cells using its invasive hyphae to enter through plasmodesmata. *M. grisea* then sporulates from the diseased rice tissue to be dispersed as conidiospores. After overwintering in sources such as rice straw and stubble, the cycle repeats.

A single cycle can be completed in about a week under favorable conditions where one lesion can generate up to thousands of spores in a single night. Disease lesions, however, can appear in three to four days after infection. With the ability to continue to produce the spores for over 20 days, rice blast lesions can be devastating to susceptible rice crops.

Environment

Rice blast is a significant problem in temperate regions and can be found in areas such as irrigated lowland and upland. Conditions conducive for rice blast include long periods of free moisture where leaf wetness is required for infection and high humidity is common. Sporulation increases with high relative humidity and at 77-82 °F, spore germination, lesion formation, and sporulation are at optimum levels.

In terms of control, excessive use of nitrogen fertilization as well as drought stress increase rice susceptibility to the pathogen as the plant is placed in a weakened state and its defenses are low. Extended drain periods also favor infection as they aerate the soil, converting ammonium to nitrate and thus causing stress to rice crops, as well.

Management

J. Sendra rice affected by Magnaporthe grisea.

The fungus has been able to establish resistance to both chemical treatments and genetic resistance in some types of rice developed by plant breeders. It is thought that the fungus can achieve this by genetic change through mutation. In order to most effectively control infection by *M. grisea*, an integrated management program should be implemented to avoid overuse of a single control method and fight against genetic resistance. For example, eliminating crop residue could reduce the occurrence of overwintering and discourage inoculation in subsequent seasons. Another strategy would be to plant resistant rice varieties that are not as susceptible to infection by

M. grisea. Knowledge of the pathogenicity of *M. grisea* and its need for free moisture suggest other control strategies such as regulated irrigation and a combination of chemical treatments with different modes of action. Managing the amount of water supplied to the crops limits spore mobility thus dampening the opportunity for infection. Chemical controls such as Carpropamid have been shown to prevent penetration of the appressoria into rice epidermal cells, leaving the grain unaffected.

VIROID

Viroids are the smallest infectious pathogens known. They are composed solely of a short strand of circular, single-stranded RNA that has no protein coating. All known viroids are inhabitants of higher plants, in which most cause diseases, ranging in economic importance.

Discovery of the viroid triggered the third major extension of the biosphere in history to include smaller lifelike entities—after the discovery of the "subvisible" microorganisms by Antonie van Leeuwenhoek in 1675 and the "submicroscopic" viruses by Dmitri Iosifovich Ivanovsky in 1892.

The unique properties of viroids have been recognized by the International Committee for Virus Taxonomy with the creation of a new order of subviral agents.

The first recognized viroid, the pathogenic agent of the potato spindle tuber disease, was discovered, initially molecularly characterized, and named by Theodor Otto Diener, plant pathologist at the U.S Department of Agriculture's Research Center in Beltsville, Maryland, in 1971. This viroid is now called Potato spindle tuber viroid, abbreviated PSTVd.

In a year 2000 compilation of the most important Millennial Milestones in Plant Pathology, the American Phytopathological Society has ranked the 1971 discovery of the viroid as one of the Millennium's ten most important pathogen discoveries.

As cogently expressed by Flores et al: "Viruses (and viroids) share the most characteristic property of living beings: In an appropriate environment, they are able to generate copies of themselves, in other words, they are endowed with autonomous replication (and evolution). It is in this framework where viroids represent the frontier of life (246 to 467nt)."

Although viroids are composed of nucleic acid, they do not code for any protein. The viroid's replication mechanism uses RNA polymerase II, a host cell enzyme normally associated with synthesis of messenger RNA from DNA, which instead catalyzes "rolling circle" synthesis of new RNA using the viroid's RNA as a template. Some viroids are ribozymes, having catalytic properties that allow self-cleavage and ligation of unit-size genomes from larger replication intermediates.

With Diener's 1989 hypothesis that viroids may represent "living relics" from the widely assumed, ancient, and non-cellular RNA world—extant before the evolution of DNA or proteins—viroids have assumed significance beyond plant pathology to evolutionary science, by representing the most plausible RNAs capable of performing crucial steps in abiogenesis, the evolution of life from inanimate matter.

Taxonomy

- Family Pospiviroidae:
    - Genus Pospiviroid; type species: Potato spindle tuber viroid; 356–361 nucleotides(nt)
    - Genus Pospiviroid; another species: Citrus exocortis viroid; 368–467 nt
    - Genus Hostuviroid; type species: Hop stunt viroid; 294–303 nt
    - Genus Cocadviroid; type species: Coconut cadang-cadang viroid; 246–247 nt
    - Genus Apscaviroid; type species: Apple scar skin viroid; 329–334 nt
    - Genus Coleviroid; type species: Coleus blumei viroid 1; 248–251 nt

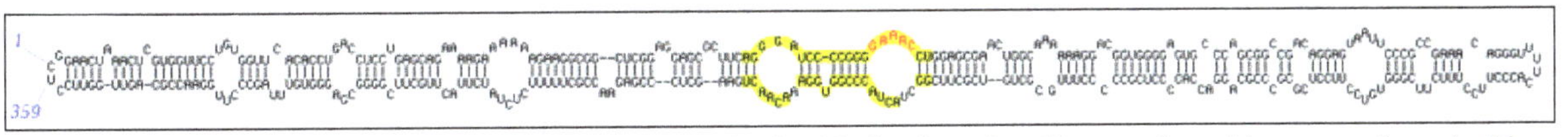

Putative secondary structure of the PSTVd viroid. The highlighted nucleotides are found in most other viroids.

- Family Avsunviroidae:
    - Genus Avsunviroid; type species: Avocado sunblotch viroid; 246–251 nt
    - Genus Pelamoviroid; type species: Peach latent mosaic viroid; 335–351 nt
    - Genus Elaviroid; type species: Eggplant latent viroid; 332–335 nt

Transmission

The reproduction mechanism of a typical viroid. Leaf contact transmits the viroid. The viroid enters the cell via its plasmodesmata. RNA polymerase II catalyzes rolling-circle synthesis of new viroids.

Viroid infections can be transmitted by aphids, by cross contamination following mechanical damage to plants as a result of horticultural or agricultural practices, or from plant to plant by leaf contact.

Replication

Viroids replicate in the nucleus (*Pospiviroidae*) or chloroplasts (*Avsunviroidae*) of plant cells in three steps through an RNA-based mechanism. They require RNA polymerase II, a host cell enzyme normally associated with synthesis of messenger RNA from DNA, which instead catalyzes "rolling circle" synthesis of new RNA using the viroid as template.

RNA Silencing

There has long been uncertainty over how viroids induce symptoms in plants without encoding any protein products within their sequences. Evidence suggests that RNA silencing is involved in the process. First, changes to the viroid genome can dramatically alter its virulence. This reflects the fact that any siRNAs produced would have less complementary base pairing with target messenger RNA. Secondly, siRNAs corresponding to sequences from viroid genomes have been isolated from infected plants. Finally, transgenic expression of the noninfectious hpRNA of potato spindle tuber viroid develops all the corresponding viroid-like symptoms. This indicates that when viroids replicate via a double stranded intermediate RNA, they are targeted by a dicer enzyme and cleaved into siRNAs that are then loaded onto the RNA-induced silencing complex. The viroid siRNAs contain sequences capable of complementary base pairing with the plant's own messenger RNAs, and induction of degradation or inhibition of translation causes the classic viroid symptoms.

RNA World Hypothesis

Diener's 1989 hypothesis had proposed that the unique properties of viroids make them more plausible macromolecules than introns, or other RNAs considered in the past as possible "living relics" of a hypothetical, pre-cellular RNA world. If so, viroids have assumed significance beyond plant virology for evolutionary theory, because their properties make them more plausible candidates than other RNAs to perform crucial steps in the evolution of life from inanimate matter (abiogenesis). Diener's hypothesis was mostly forgotten until 2014, when it was resurrected in a review article by Flores et al. in which the authors summarized Diener's evidence supporting his hypothesis as:

- Viroids' small size, imposed by error-prone replication.
- Their high guanine and cytosine content, which increases stability and replication fidelity.
- Their circular structure, which assures complete replication without genomic tags.
- Existence of structural periodicity, which permits modular assembly into enlarged genomes.
- Their lack of protein-coding ability, consistent with a ribosome-free habitat.
- Replication mediated in some by ribozymes—the fingerprint of the RNA world.

The presence, in extant cells, of RNAs with molecular properties predicted for RNAs of the RNA World constitutes another powerful argument supporting the RNA World hypothesis.

PHYTOPLASMA

Phytoplasmas are obligate bacterial parasites of plant phloem tissue and of the insect vectors that are involved in their plant-to-plant transmission. Phytoplasmas were discovered in 1967 by Japanese scientists who termed them mycoplasma-like organisms. Since their discovery, phytoplasmas have resisted all attempts at *in vitro* culture in any cell-free medium; routine cultivation in an artificial medium thus remains a major challenge. Although phytoplasmas have recently been reported to be grown in a specific artificial medium, experimental repetition has yet to be reported. Phytoplasmas are characterized by the lack of a cell wall, a pleiomorphic or filamentous shape, a diameter normally less than 1 μm, and a very small genome.

Phytoplasmas are pathogens of agriculturally important plants, including coconut, sugarcane, and sandalwood, in which they cause a wide variety of symptoms ranging from mild yellowing to death. Phytoplasmas are most prevalent in tropical and subtropical regions. They are transmitted from plant to plant by vectors (normally sap-sucking insects such as leafhoppers) in which they both survive and replicate.

References to diseases now known to be caused by phytoplasmas can be found as far back as 1603 (mulberry dwarf disease in Japan.) Such diseases were originally thought to be caused by viruses, which, like phytoplasmas, require insect vectors, and cannot be cultured. Viral and phytoplasmic infections share some symptoms. In 1967, phytoplasmas were discovered in ultrathin sections of plant phloem tissue and were termed mycoplasma-like organisms due to their physiological resemblance The organisms were renamed phytoplasmas in 1994, at the 10th Congress of the International Organization for Mycoplasmology.

Morphology

Phytoplasmas are Mollicutes, which are bound by a triple-layered membrane, rather than a cell wall. The phytoplasma cell membranes studied to date usually contain a single immunodominant protein of unknown function that constitutes most of the protein in the membrane. A typical phytoplasma is pleiomorphic or filamentous in shape and is less than 1 μm in diameter. Like other prokaryotes, phytoplasmic DNA is distributed throughout the cytoplasm, instead of being concentrated in a nucleus.

Symptoms

Phytoplasmas can infect and cause various symptoms in more than 700 plant species. One characteristic symptom is abnormal floral organ development including phyllody, (i.e. the production of leaf-like structures in place of flowers) and virescence (i.e. the development of green flowers attributable to a loss of pigment by petal cells). Phytoplasma-harboring flowering plants may nevertheless be sterile. The expression of genes involved in maintaining the apical meristem or in the development of floral organs is altered in the morphologically affected floral organs of phytoplasma-infected plants.

A phytoplasma infection often triggers leaf yellowing, probably due to the presence of phytoplasma cells in phloem, which can affect phloem function and carbohydrate transport, inhibit chlorophyll biosynthesis, and trigger chlorophyll breakdown. These symptoms may be attributable to stress caused by the infection rather than a specific pathogenetic process.

Many phytoplasma-infected plants develop a bushy or "witch's broom" appearance due to changes in their normal growth patterns. Most plants exhibit apical dominance but infection can trigger the proliferation of axillary (side) shoots and a reduction in internode size. Such symptoms are actually useful in the commercial production of poinsettias. Infection triggers more axillary shoot production; the poinsettia plants thus produce more than a single flower.

Effector (Virulence) Proteins

Many plant pathogens produce virulence factors (i.e. effectors) that modulate or interfere with normal host processes to the benefit of the pathogens. In 2009, a secreted protein, termed "tengu-su inducer" (TENGU; CoH5W6), was identified from a phytoplasma causing yellowing of onions; this was the first phytoplasmal virulence factor to be described. TENGU induces characteristic symptoms, including witches' broom and dwarfism. Transgenic expression of TENGU in *Arabidopsis* plants induced sterility in male and female flowers. TENGU contains a signal peptide at its N-terminus; after cleavage, the mature protein is only 38 amino acids in length. Although phytoplasmas are restricted to phloem, TENGU is transported from phloem to other cells, including those of the apical and axillary meristems. TENGU was suggested to inhibit both auxin- and jasmonic acid-related pathways, thereby affecting plant development. Surprisingly, the N-terminal 11 amino acid region of the mature protein triggers symptom development in *Nicotiana benthamiana* plants. TENGU undergoes proteolytic processing by a plant serine protease *in vivo*, suggesting that the N-terminal peptide (i.e. the 11 amino acid fragment) alone induces the observed symptoms. TENGU homologs have been identified in AY-group phytoplasmas. All such homologs undergo processing and can induce symptoms, suggesting that the symptom-inducing mechanism is conserved among TENGU homologs.

In 2009, 56 genes for secreted proteins were identified in the genome of Aster Yellows phytoplasma strain Witches Broom (AY-WB); these were named secreted AY-WB proteins (SAPs) and considered effectors. Also in 2009, effector SAP11 was shown to target plant cell nuclei and unload from phloem cells in AY-WB-infected plants. SAP11 was found to induce stem proliferations and changes of leaf shapes of plants; the stem proliferations induced by SAP11 resemble witch's broom symptoms of AY-WB-infected plants. In addition, it was demonstrated that SAP11 interacts with and destabilizes plant class II TCP protein domain transcription factors that leads to shoot proliferations and leaf shape changes. In addition to regulation of plant development, TCPs also control the expression of lipoxygenase genes required for jasmonate biosynthesis. Jasmonate levels are decreased in phytoplasma-infected *Arabidopsis* plants and plants that transgenically express the AY-WB SAP11 effector. The downregulation of jasmonate production is beneficial to phytoplasmas because jasmonate is involved in plant defenses against herbivorous insects such as leafhoppers. Leafhoppers lay increased numbers of eggs on AY-WB-infected plants, at least in part because of SAP11 production. For example, the leafhopper *Macrosteles quadrilineatus* laid 30% more eggs on plants that expressing SAP11 transgenically than control plants, and 60% more eggs on plants infected with AY-WB. Phytoplasmas cannot survive in the external environment and are dependent upon insects such as leafhoppers for transmission to new (healthy) plants. Thus, by compromising jasmonate production, SAP11 encourages leafhoppers to lay more eggs on phytoplasma-infected plants, thereby ensuring that newly hatched leafhopper nymphs feed upon infected plants to become phytoplasma vectors. SAP11 effectors are identified in a number of divergent phytoplasmas and these effectors also interact with TCPs and modulate plant defenses. SAP11 is the first phytoplasma virulence protein for which plant targets and effector functions (i.e. evidence of benefit for the pathogen) were identified. TCPs were found to be targeted by a number of other pathogen effectors.

The AY-WB phytoplasma effector SAP54 was shown to induce virescence and phyllody when expressed in plants and homologs of this effector were found in at least three other phytoplasmas. Two SAP54 homologs, PHYL1 of the onion yellows phytoplasma and $PHYL1_{PnWB}$ of the peanut witches' broom phytoplasma, also induce phyllody-like floral abnormalities. These results suggest that PHYL1, SAP54, and their homologs form a phyllody-inducing gene family, the members of which are termed phyllogens. MADS-box transcription factors (MTFs) of the ABCE model play critical roles in floral organ development in *Arabidopsis*. Phyllogens interact directly with class A and class E MTFs, inducing protein degradation in a ubiquitin/proteasome-dependent manner that, at least for SAP54, is dependent on interactions with the proteasome shuttle factor RAD23. Interestingly, RAD23 mutants do not show phyllody when infected with phytoplasma indicating that RAD23 proteins are susceptibility factors; i.e. phytoplasmas and SAP54 require these plant proteins to induce phyllody symptoms. The accumulation of mRNAs encoding class B MTFs, the transcription of which is positively regulated by class A and class E MTFs, is drastically decreased in Arabidopsis constitutively expressing PHYL1. Phyllogens induce abnormal floral organ development by inhibiting the functions of these MTFs. RAD23 proteins are also required for promoting leafhopper vector egg laying on plants that express SAP54 and are infected with AY-WB phytoplasma.

Transmission

Movement between Plants

Phytoplasmas are spread principally by insects of the families Cicadellidae (leafhoppers), Fulgoridae (planthoppers), and Psyllidae (jumping plant lice), which feed on the phloem of infected plants, ingesting phytoplasmas and transmitting them to the next plant on which they feed. Thus, the host range of phytoplasmas is strongly dependent upon that of the insect vector. Phytoplasmas contain a major antigenic protein constituting most of the cell surface protein. This protein associates with insect microfilament complexes and is believed to control insect-phytoplasma interactions. Phytoplasmas can overwinter in insect vectors or perennial plants. Phytoplasmas can have varying effects on their insect hosts; examples of both reduced and increased fitness have been noted.

Phytoplasmas enter the insect body through the stylet, pass through the intestine, and then move to the hemolymph and colonize the salivary glands: the entire process can take up to 3 weeks. Once established in an insect host, phytoplasmas are found in most major organs. The time between ingestion by the insect and attainment of an infectious titer in the salivary glands is termed the latency period.

Phytoplasmas can also be spread via dodders (*Cuscuta*) or by vegetative propagation such as the grafting of infected plant tissue onto a healthy plant.

Movement within Plants

Phytoplasmas move within phloem from a source to a sink, and can pass through sieve tube element. However, as phytoplasmas spread more slowly than solutes, and for other reasons, passive translocation within plants is thought to be unimportant.

Detection and Diagnosis

Before the molecular era, the diagnosis of phytoplasma-caused diseases was difficult because the organisms could not be cultured. Thus, classical diagnostic techniques, including symptom observation were used. Ultrathin sections of phloem tissue from plants with suspected phytoplasma-infections were also studied. The empirical use of antibiotics such as tetracycline was additionally employed.

Molecular diagnostic techniques for phytoplasma detection began to emerge in the 1980s and included enzyme-linked immunosorbent assay (ELISA)-based methods. In the early 1990s, polymerase chain reaction (PCR)-based techniques were developed: these are far more sensitive than ELISAs, and restriction fragment length polymorphism (RFLP) analysis allowed the accurate identification of various phytoplasma strains and species.

More recent techniques allow infection levels to be assessed. Both quantitative PCR and bioimaging can effectively quantify phytoplasma titers within plant. In addition, loop-mediated isothermal amplification (a sensitive, simple, and rapid diagnostic method) is now available as a commercial kit allowing all known phytoplasma species to be detected in about 1 h, including the DNA extraction step.

Control

Phytoplasmas are normally controlled by the breeding and planting of disease-resistant crop varieties (perhaps the most economically viable option) and by the control of insect vectors.

Tissue culture can be used to produce healthy clones of phytoplasma-infected plants. Cryotherapy (i.e. the freezing of plant samples in liquid nitrogen) prior to tissue culture increases the probability of producing healthy plants in this manner.

Plantibodies targeting phytoplasmas have also been developed.

Tetracyclines are bacteriostatic to phytoplasmas. However, disease symptoms reappear in the absence of continuous antibiotic application. Thus, tetracycline is not a viable agricultural control agent, but it is used to protect ornamental coconut trees.

Genetics

The genomes of four phytoplasmas have been sequenced: "onion yellows", "aster yellows witches' broom", *Ca.* Phytoplasma australiense, and Ca. Phytoplasma Mali. Phytoplasmas have very small genomes, with extremely small amount of G and C nucleotides (sometimes as little as 23%, which is thought to be the lower threshold for a viable genome). In fact, the Bermuda grass white-leaf phytoplasma has a genome size of only 530 kb, one of the smallest known genomes of all living organisms. The larger phytoplasma genomes are around 1350 kb in size. The small genome size of phytoplasma is attributable to reductive evolution from *Bacillus/Clostridium* ancestors. Phytoplasmas have lost ≥75% of their original genes, and can thus no longer survive outside of insects or plant phloem. Some phytoplasmas contain extrachromosomal DNA such as plasmids.

Despite their small genomes, many predicted phytoplasma genes are present in multiple copies. Phytoplasmas lack many genes encoding standard metabolic functions and have no functioning homologous recombination pathway, but they do have a *sec* transport pathway. Many phytoplasmas contain two rRNA operons. Unlike other Mollicutes, the triplet code of UGA is used as a stop codon in phytoplasmas.

Phytoplasma genomes contain large numbers of transposons and insertion sequences and also contain a unique family of repetitive extragenic palindromes termed PhREPS for which no role is known. However, it is theorized that the stem-loop structures in PhREPS play a role in transcription termination or genome stability.

Taxonomy

Phytoplasmas belong to the monotypic order Acholeplasmatales. In 1992, the Subcommittee on the Taxonomy of Mollicutes proposed the use of "*Phytoplasma*" rather than "mycoplasma-like organisms" "for reference to the phytopathogenic mollicutes". In 2004, the generic name phytoplasma was adopted and is currently of Candidatus (Ca.) status (used for bacteria that cannot be cultured). Phytoplasma taxonomy is complicated because the organisms cannot be cultured; methods normally used to classify prokaryotes are thus not available. Phytoplasma taxonomic groups are based on differences in fragment sizes produced by restriction digests of 16S ribosomal RNA gene sequences (RFLPs) or by comparisons of DNA sequences from 16s/23s spacer regions. The actual number of taxonomic groups remains unclear; recent work on computer-simulated restriction digests of the 16Sr gene suggested up to 28 groups, whereas others have proposed fewer groups, but more subgroups. Each group includes at least one *Ca.* Phytoplasma species, characterized by distinctive biological, phytopathological, and genetic properties.

Symptoms of aster yellows on marigold.

A grape vine with "bois noir" phytoplasma disease.

Coconut palms dying of lethal yellowing disease.

OOMYCETES

The oomycetes, also known as "water molds", are a group of several hundred organisms that include some of the most devastating plant pathogens. The diseases they cause include seedling blights, damping-off, root rots, foliar blights and downy mildews. Some notable diseases are the late blight of potato, downy mildew of grape vine, sudden oak death, and root and stem rot of soybean. Because of their filamentous growth habit, nutrition by absorption, and reproduction via spores, oomycetes were long regarded by plant pathologists as lower fungi. However, as our understanding of evolutionary relationships has grown, it is now clear that this group of organisms is unrelated to the true fungi. Indeed, fungi appear more closely related to animals than to oomycetes, and oomycetes are more closely related to algae and to green plants.

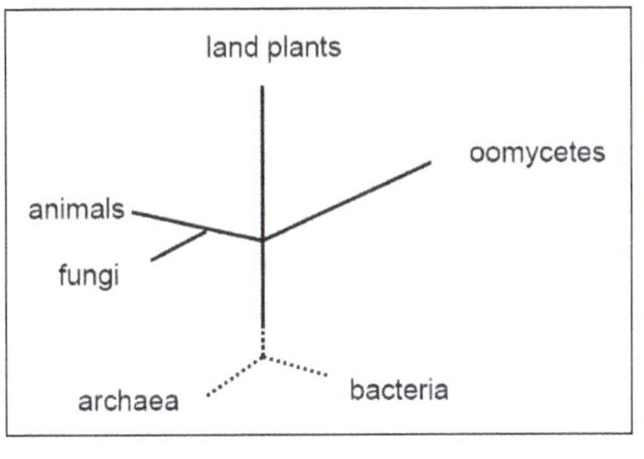

Phylogenetic analyses using genes and intergenic regions have confirmed the assertions of earlier systematists that the oomycetes are different from fungi. The data from these molecular analyses have been particularly convincing to non-systematist plant pathologists. There are many features distinguishing oomycetes from fungi. Septa (cell walls) in the hyphae are rare, resulting in a multinucleate condition (termed coenocytic). The nuclei of vegetative cells are typically diploid. The cell wall is composed of β-1,3, and β-1,6 glucans, and not of chitin (the polymer of N-acetyl glucose amine, found in the walls of true fungi). Many species produce wall-less, biflagellated swimming spores (zoospores) in structures called sporangia.

Morphological Characteristics of Oomycetes

One of the most distinguishing characteristics is the production of zoospores produced in sporangia. The anterior flagellum of a zoospore is a tinsel type, while the posterior flagellum is a whiplash type; both are typically attached in a ventral groove. Although wall-less, zoospores retain a consistent but flexible shape. Zoospores can swim in water films on leaf surfaces, in soil water, in hydroponic media and in natural bodies of water. Oomycetes can often be "baited" from soil water, streams or ponds, and it is thought that zoospores are attracted to the baits. After a time of free swimming the zoospores settle on a surface, retract their flagella, and secrete a mucilaginous matrix which affixes them to the surface. Sporangia of different taxa within the group are of diverse shapes and characteristics. They may be terminal or intercalary (within a hyphal filament), bulbous or not, and if terminal, caducous (sporangia detach readily) or not.

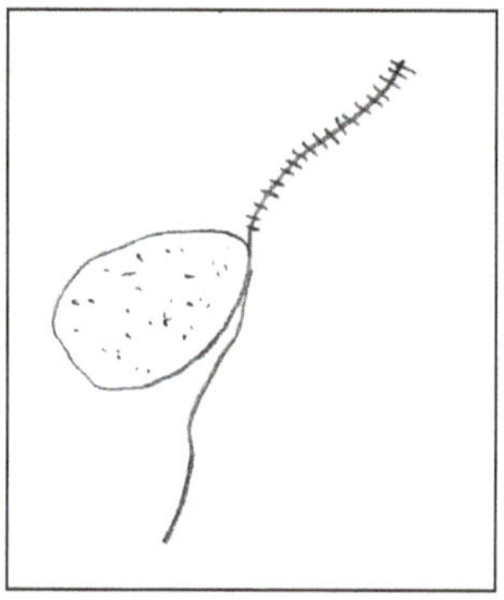

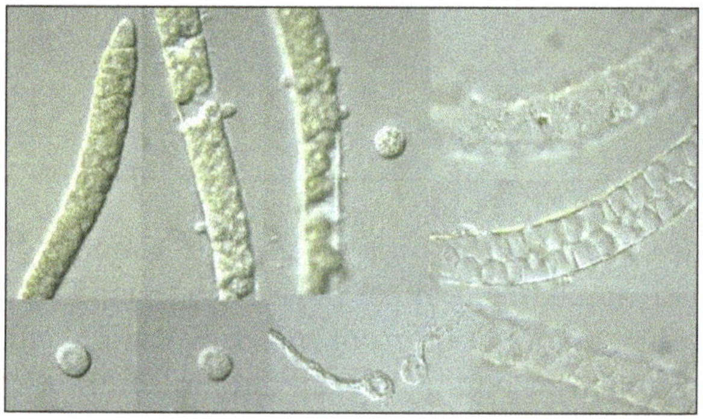

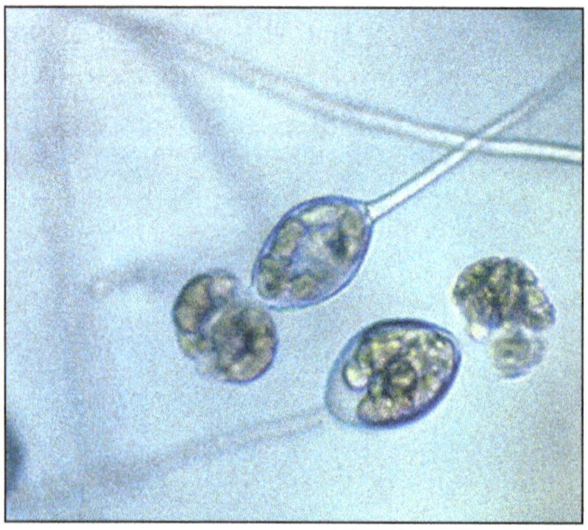

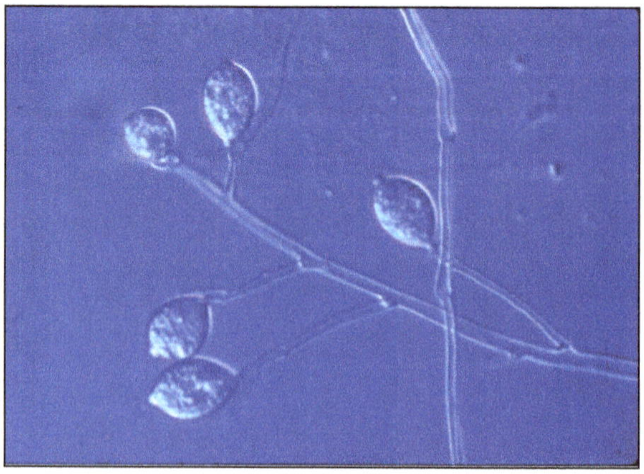

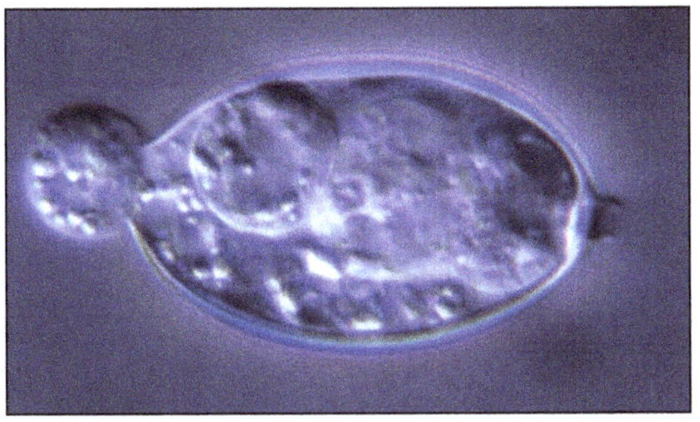

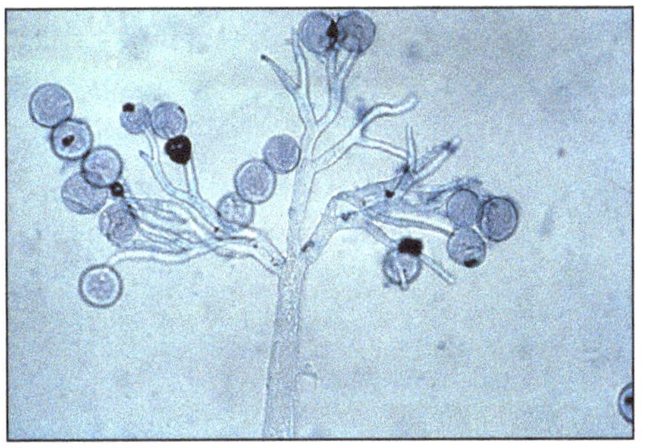

In some species, the ability to produce zoospores has been lost, and sporangia are thought to have evolved into structures that germinate directly to produce germ tubes. In this case, the sporangia are sometimes termed "conidia". In yet other species, sporangia can germinate directly to produce germ tubes or "indirectly" to produce zoospores, a trait which is often temperature dependent, with zoospores being produced at cooler temperatures.

Sexual reproduction occurs via the production of gametangia: oogonia and antheridia. Because meiosis does not occur until the formation of gametangia occurs, the vegetative nuclei are diploid. The morphology of antheridium attachment has been an important feature in morphological taxonomy of some genera. In some genera the antheridium is attached to the side of the oogonium, but in other genera, the antheridium surrounds the base of the oogonium.

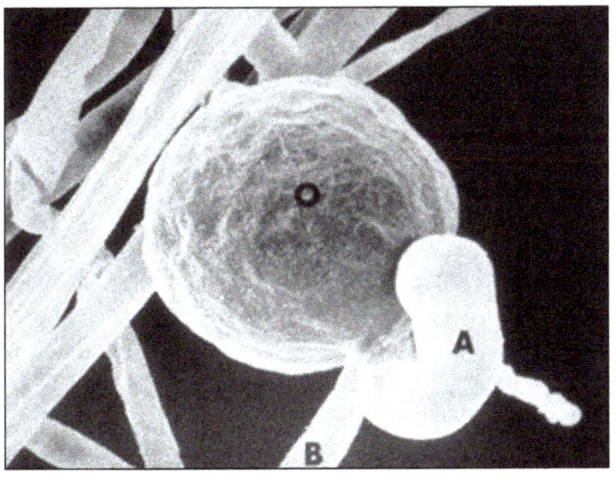

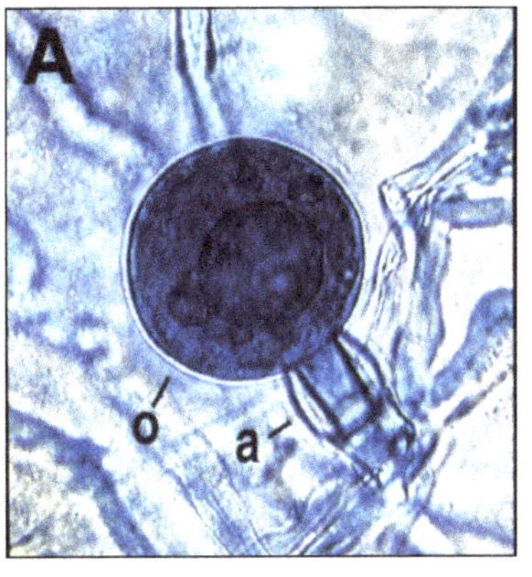

Typically each individual produces both antheridia and oogonia. There may be differences in "femaleness" and "maleness" and sexual preference is relative to other individuals. In some species, two distinct mating types occur and both are required for sexual reproduction (these are heterothallic as opposed to homothallic species). In heterothallic oomycetes, the gametangia are produced only in the presence of both mating types due to the fact that a hormone produced by one thallus stimulates the other to produce gametangia. In other species, sexual reproduction occurs within a single individual (these are homothallic individuals). Unlike the heterothallic species, homothallic individuals do not require distinct mating types, but can reproduce sexually by selfing. All Pythium and some Phytophthora species are homothallic.

The fertilized oogonium develops into a thick-walled oospore. When the oospores are produced in plant tissue, they may occupy a large portion of the tissue. Oospores of many species have been shown to be able to survive for years in soil.

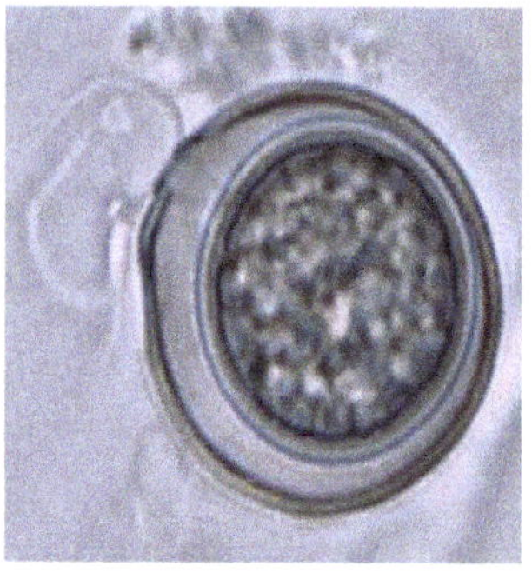

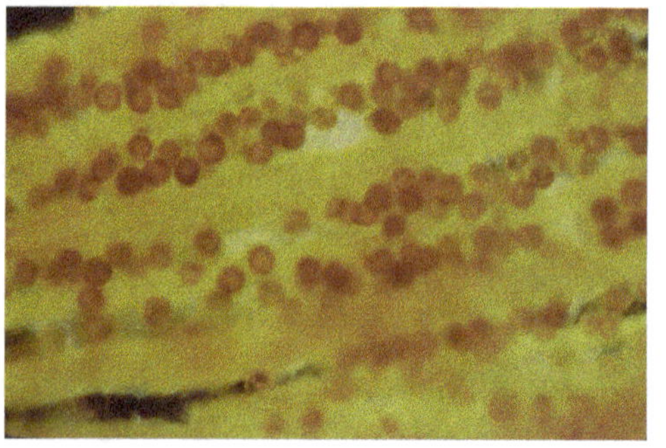

After a period of dormancy (often of apparently diverse and undefined durations) oospores germinate to produce hyphae, which may immediately produce a sporangium. Oospore germination is often asynchronous; that is, some oospores germinate while others do not. Germination and survival of oospores is dependent on environmental conditions: generally, oospores are able to survive dry and cool or cold conditions, but seem sensitive to high temperatures (> 40-45 °C).

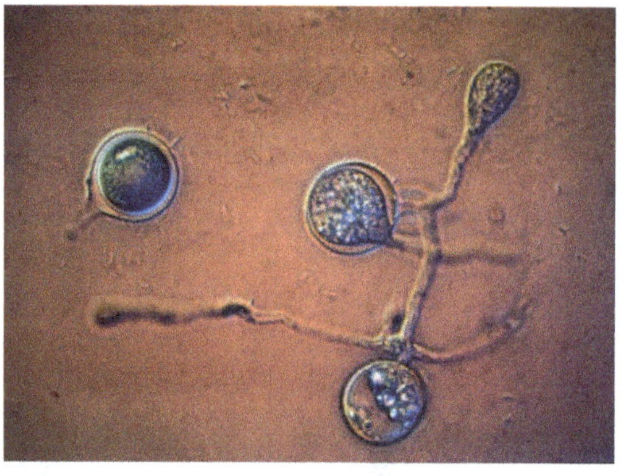

Some species produce thick-walled survival structures called chlamydospores.

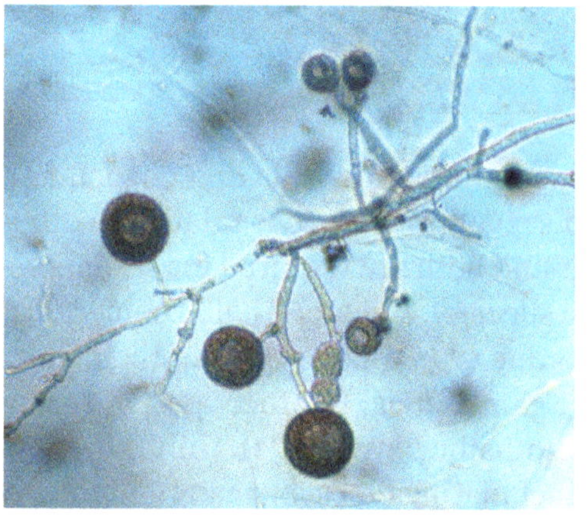

Relationships within the Oomycetes

Our understanding of the relationships among oomycetes is evolving rapidly as we gather additional information, particularly from molecular analyses. The techniques have evolved rapidly and analysis of DNA sequence provides a common criterion for assessing relationships. The analysis of probable relationships among the major genera of oomycetes is depicted in Figure. While Pythium, Phytophthora and Peronospora appear related, the relationship of these organisms with the grass downy mildews remains problematic.

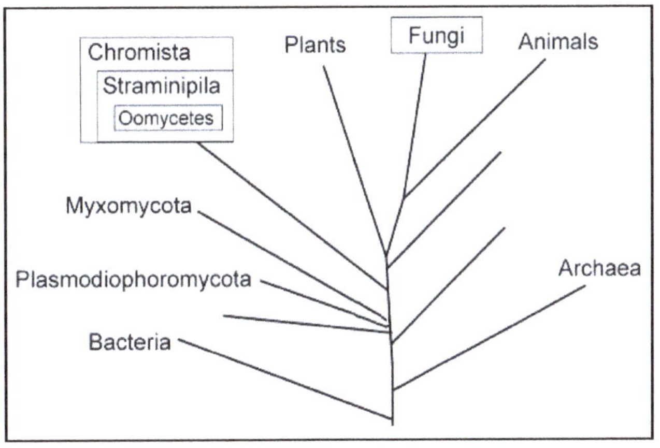

Oomycete Plant Pathogens

Phytophthora Infestans

This is the pathogen that caused the Irish potato famine in the mid-19th century. It was first reported in the eastern United States just prior to reports of its presence in Europe. Prior to that time, it was not known to western science. However, its devastating impact on potatoes and the terrible misery it has caused have made it infamous. Foliage, stems and tubers are susceptible. It is a heterothallic species, with only one mating type (the A1) historically dominating the worldwide population with the exception of populations that existed in Mexico; both mating types have existed in central Mexico for a very long time. However, in the late 20th century migrations from Mexico distributed a very complex and diverse population containing both A1 and A2 mating types to Europe: subpopulations were later distributed from Europe to other locations. As an asexual organism in nature or in agriculture, P. infestans is essentially an obligate parasite, with no long-term survival mechanism; potato tubers provide a mechanism for short-term survival if infected tubers are stored between cropping seasons. However, the relatively recent migrations of the A1 and A2 mating types increases the chances of sexual reproduction and the production of oospores that would represent a long-term survival mechanism for this devastating pathogen.

Phytophthora infestans is unusual for a Phytophthora in that it is an aerial pathogen. That is, it infects and reproduces mainly on the above ground portions of its host. Sporangia are dehiscent (detach easily when mature) and under cloudy conditions can survive transit sufficiently long to travel many kilometers in moving bodies of air. The pathogen is favored by moist, cool environments: sporulation is optimal at 12-18° C in water-saturated or nearly saturated environments, and zoospore production is favored at temperatures below 15° C. Lesion growth rates are typically optimal at a slightly warmer temperature range of 20 to 24° C. Under favorable conditions, the asexual life cycle (sporangium germination, infection, lesion growth, sporulation) can be completed within as few as four days, but symptoms may not be visible for the first 2-3 days after initial infection. The dominant influence of weather on the infection and sporulation process of P. infestans has caused investigators to develop various forecasts for late blight. These investigations have resulted in algorithms (Dutch Rules, Beaumont periods, rain favorable days, severity values, etc.), which identify weather that has been favorable for late blight and allow growers to predict conditions that are likely to encourage or enhance infection. The explosive potential of this pathogen is legendary, dramatic and real. When the disease is uncontrolled and when environmental conditions are favorable to the pathogen, fields of 10-40 acres will succumb to the disease within just a few days.

The general susceptibility of potatoes and tomatoes has stimulated much effort to develop resistant plants as well as to understand the pathogenicity of P. infestans. Single large-effect genes for resistance (R genes) have been identified and deployed. Unfortunately, because of variation in the pathogen population, the effect of these genes has not been long lasting. R genes recognize specific components of pathogen proteins (effectors) that are injected into the host cell. Mutation in these effectors can enable the pathogen to escape recognition and avoid the resistance mechanisms. There have also been efforts made to create resistant plants based on a less well-understood mechanism that may involve many genes. This mechanism has been termed "field" or "partial" resistance. However, the most popular cultivars of potatoes and tomatoes are quite susceptible, necessitating the use of fungicides to protect plants.

Effective disease suppression requires a strategy integrating several tactics. Because infected seed tubers can be a source of the pathogen, it is important to plant only healthy seed tubers. It is also important to eliminate any tubers that might have survived from one cropping season to the next, whether these tubers survived in soil after harvest or were discarded after storage. Some Solanaceous weeds can also harbor the pathogen, and any infected weeds in or near the crop need to be eliminated. Fungicides are used in connection with a good scouting (monitoring) program (to learn if the pathogen is present) and application timing and rates are often aided by an appropriate forecast.

Plasmopara Viticola

This pathogen was introduced to Europe from North America in the late 19th century. It accompanied wild grape plants imported for their resistance to the sap-sucking insect pest Phylloxera. P. viticola is a heterothallic downy mildew with A1 and A2 mating types. Oospores germinate to produce sporangia with zoospores, which can be splash-dispersed to cause lesions. Sporangia from primary lesions can also be wind-dispersed. Symptoms on leaves are small yellow lesions also known as oil spots. European grape varieties are susceptible to P. viticola and fungicides are used extensively to suppress the disease. Forecast systems are used to improve the efficiency of disease suppression.

Phytophthora Cinnamomi

This devastating, omnivorous pathogen was first isolated in the early 20th century, and is thought by some to have originated in Papua New Guinea, but it now has a worldwide distribution. Its host range is thought to include more than 3000 species of plants. It is heterothallic with A1 and A2 mating types, but sexual recombination is not thought to have a significant role in population diversity. Often, populations consist of a single mating type. This pathogen infects fibrous roots and can also survive and grow saprophytically in soil. It produces chlamydospores so that even in the absence of sexual reproduction, it can survive for long periods in soil.

The asexual cycle can be very rapid during wet conditions and is described elegantly by Hardham; only a summary is presented here. Sporangia are produced on sporangiophores and the sporangia release 20-30 zoospores. Zoospore formation is triggered by a decrease in temperature, resulting in the change of expression of a large number of genes. Morphologically, the cytoplasm becomes delimited into uninucleate compartments, with membranes forming around each and with each developing flagella and a water expulsion vacuole. In P. cinnamomi, but not in all species of Phytophthora, the wall material at the apex of the sporangium expands into an extra-sporangial vesicle into which the zoospores are released. The vesicle is ephemeral and the zoospores are quickly released into the environment. The zoospores may travel distances of several centimeters and are attracted to potential infection sites where they encyst. A mucilaginous material is secreted over the surface and they become affixed to the host surface (encyst) within minutes upon arrival and quickly form a cell wall. Germination occurs rapidly after encystment and the germ tubes penetrate the root epidermis. Colonization of host tissue follows and in susceptible tissues, sporulation may occur within three days.

This pathogen is a threat to both agricultural and native plants. One of the most seriously affected areas is in Australia where the pathogen was introduced in the early 20th century. The pathogen was apparently introduced with imported plants and escaped into the native eucalyptus forest, the jarrah, and the disease has thus been termed “Jarrah Dieback”. In addition to Eucalyptus, many other native species are also susceptible to this pathogen and the disease remains severe to the present time. Unfortunately, construction of logging roads has proven to be a mechanism for transport of this pathogen throughout the forest. Disease is most severe under wet and warm conditions, and disease abatement is associated with cooler drier weather. However, investigators are expecting the range of this pathogen in the northern hemisphere to extend further north in response to global climate changes. For agricultural plants, phosphonate fungicides (effective against oomycetes, but ineffective against fungi) have been particularly helpful to suppress the disease and are used extensively. Host resistance has also been investigated with some limited success.

Phytophthora Ramorum

P. ramorum is the pathogen best known for causing sudden oak death on different oak species. It also causes Ramorum blight and shoot dieback on many ornamental plants. The disease was simultaneously discovered in Europe and in California in the 1990s affecting oak and nursery crops such as rhododendron and viburnum. In the United States the sudden oak death disease gained instant notoriety because it led to extensive death in coast live oak and tan oak in California.

One aspect of this pathogen that differentiates it from many other oomycetes is that P. ramorum has a very large and diverse host range that includes many oaks, shade trees, conifers, and woody ornamentals. Symptoms differ on various hosts and can be confused with diseases caused by other organisms or even environmental factors.

Although a center of origin has not been found, scientists currently believe that P. ramorum was introduced into North America and Europe from elsewhere on imported nursery plants.

In US nurseries this pathogen is currently managed through quarantine, eradication and exclusion: nurseries are trying to avoid establishment of the pathogen by excluding it from their operations and the Oregon Department of Forestry is trying to eradicate P. ramorum in Curry County in Southern Oregon where it is entrenched in native tanoak and bay laurel (myrtle wood).

Sclerophthora Rayssiae Var. Zeae

This downy mildew is so important that it is one of the few plant pathogens to be included on the USDA-APHIS select agent list. The list is composed of plant pathogens not present in the US, but deemed to be a potential bioterrorism threat to US agriculture. Sclerophthora rayssiae var. zeae was first reported in India, but has now also been reported in Myanmar, Nepal, Pakistan, and India; the disease has been most severe in areas of high rainfall (100-200 cm/yr). In India, annual losses of 20-90% have been reported. Survival of this pathogen is via oospores in infected seeds and soil or plant debris.

The disease cycle involves both sexual and asexual reproduction. Oospores germinate to produce sporangia, which then release zoospores that penetrate leaf tissue. Lesions are initially interveinal and appear as chlorotic, brownish or reddish stripes on the leaves. Asexual sporulation is favored by moderate temperatures (20-25° C) and periods of high moisture; sporangia are produced on non-necrotic leaf tissue and give the leaf a grayish-white appearance. Sporangia are dispersed short distances via wind or rain splash, and germinate to produce zoospores or, less commonly, to produce a germ tube to repeat the cycle. Oospores are produced in necrotic tissue and can survive for years in soil or in plant debris.

Peronosclerospora Philippinensis

Due to its devastating nature and the fact that it is not yet present in the US, this pathogen has also been placed on the USDA-APHIS select agent list. Peronosclerospora philippinensis is endemic to the Philippines where annual losses of 40-60% have been reported. This oomycete does not produce zoospores, but rather the sporangia germinate directly and have been referred to as conidia. Initial infections of roots are thought to result from oospores in soil. Most infections are initiated by sporangia (conidia) produced from infected foliage that can be distributed to other plants where they germinate directly and initiate local lesions. Young plants and seeds may be infected systemically. In addition to maize, the hosts include sugar cane and other grasses, but yield losses on these other hosts are not well defined. Disease severity is highest in tropical climates and areas that receive 100-200 cm of rain annually. Epidemics occur due to the rapid secondary cycles that are driven by high moisture and warm temperatures (20-25 °C). The role of the oospore in the disease cycle has not been determined.

Pythium Aphanidermatum and P. Ultimum

Pythium species are best known for causing damping-off and seed rot disease that often occurs just after planting as young seedlings emerge. Pythium also causes root rots on newly emerged or more mature plants and can also cause soft rots of fleshy fruit.

Damping-off disease affects seedlings worldwide. Often, young seedlings are completely destroyed by this pathogen and a crop emerges unevenly, leading to significant yield reductions. Older plants once emerged might not be significantly affected by Pythium, but do show symptoms of root rot.

Several species of Pythium cause damping-off.

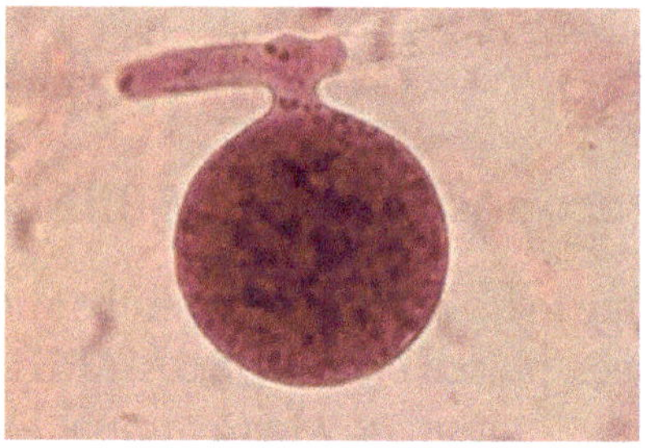

Aphanomyces Euteiches

Aphanomyces euteiches causes seedling and root-rot diseases on many legumes and is considered to be the most yield limiting pathogen of pea in some growing areas of the world. The genus Aphanomyces is particularly interesting because it includes plant and animal pathogens found in both terrestrial and aquatic habitats. A. euteiches affects a variety of legumes including alfalfa, clover, dry bean, lentil, faba bean, pea, snap bean, and several weed species. This pathogen infects the cortex of primary and lateral roots. Infected areas initially turn honey-brown; as disease progresses the cortex sloughs off and roots turn dark brown to black. Microscopic examination often reveals oospores in the cortex.

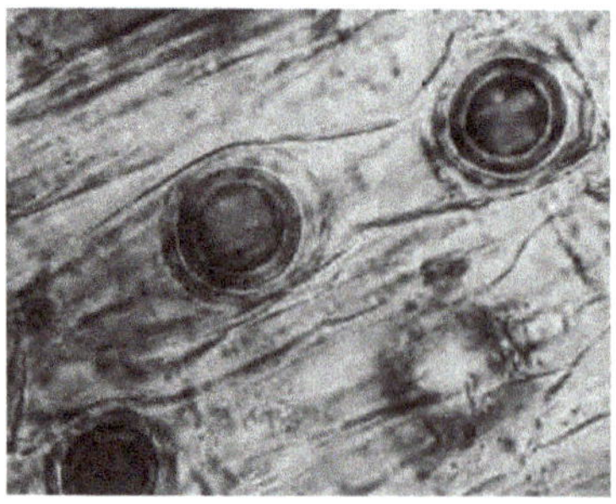
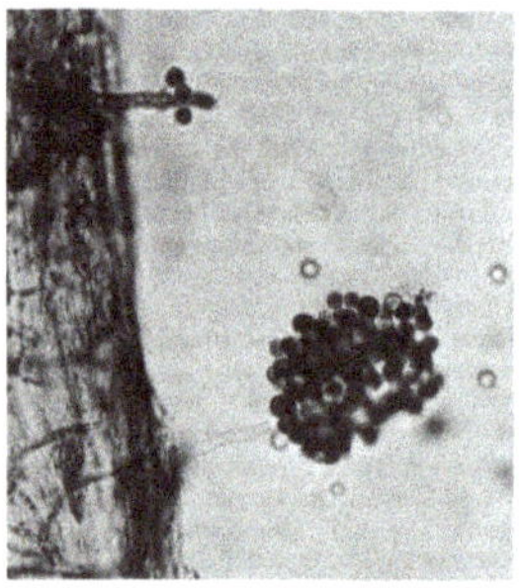

Like Pythium spp. this pathogen is able to reproduce sexually as a homothallic oomycete by producing oospores. During asexual reproduction, the pathogen produces distinct sporangia that differentiate zoospores.

References

- Plpath-gen-7, factsheet: ohioline.osu.edu, Retrieved 31 March, 2019

- Roossinck, M. J. (2011). “The good viruses: viral mutualistic symbioses”. Nature Reviews Microbiology. 9 (2): 99–108. doi:10.1038/nrmicro2491. PMID 21200397
- Diseases, professional: cyclamen.com, Retrieved 14 July, 2019
- Stienstra, T. (11 October 2007). “It’s no snow job - handful of redwoods are rare albinos”. San Francisco Chronicle. Retrieved December 6, 2010
- Heide-Jørgensen, Henning (2008-06-19). Parasitic flowering plants. BRILL. doi:10.1163/ej.9789004167506.i-438. ISBN 9789047433590
- Talbot, N. J. (2003). “ON THE TRAIL OF A CEREAL KILLER: Exploring the Biology of Magnaporthe grisea”. Annual Review of Microbiology. 57: 177–202
- Alves C, Branco C, Cunha C (2013). “Hepatitis delta virus: a peculiar virus”. Adv Virol. 2013: 560105. doi:10.1155/2013/560105. PMC 3807834. PMID 24198831

3

Abiotic Plant Pathogens

Abiotic plant pathogens include a number of soil-born plant pathogens such as phytophthora cinnamomi, gaeumannomyces graminis and agrobacterium tumefaciens. Other factors such as nutrient deficiency, extreme temperatures and toxic chemicals like herbicides, fungicides and air pollutants can also cause plant diseases. This chapter has been carefully written to provide an easy understanding of these abiotic plant pathogens.

Plants can be damaged by noninfectious factors, causing problems that can collectively be termed "abiotic diseases" or "abiotic disorders". Unfavorable soil properties, fertility imbalances, moisture extremes, temperature extremes, chemical toxicity, physical injuries, and other problems are examples of abiotic disorders that can reduce plant health and even kill plants. Furthermore, many of these abiotic disorders can predispose plants to diseases caused by infectious microbes.

Diseases caused by abiotic or non-pathogenic factors cannot be transmitted from affected plants to healthy plants. Some abiotic diseases are the result of genetic changes that occur in the meristematic cells of plants, resulting in a flower, fruit or branch differing in appearance from those on the rest of the plant. The extent of the disorder depends on how early in the plant's development the change occurs. Citrus plants are particularly prone to such changes which are referred to as chimeras. Leaves on affected twigs may become variegated, fruit may have corrugated or differently-coloured sectors, branches may develop a willowy growth habit, stems may be twisted or flattened or galls or rough bark may be produced on the stems. If the cell change takes place at a very early stage of development, the whole plant may be affected. Generally these changes are undesirable, but occasionally a mutant with desirable properties such as early or late maturing fruit, high productivity or special fruit characteristics occurs. A new, improved cultivar may result. The Washington Navel orange cultivar, many strains of Red Delicious apple as well as other fruits such as plum, peach, nectarine and European pear have arisen in this way.

More often, abiotic diseases are caused by unfavourable environmental conditions. For example, a lack, excess, imbalance or deleterious interaction of the physical or chemical factors necessary for healthy plant growth may cause abnormal development. Environmental factors that affect plant growth include temperature, moisture, light and the properties of the medium in which the plant is growing. For each environmental factor, there is an optimum range within which plants grow best. Plants can tolerate some variations either side of the optimum for each environmental factor. However, they may exhibit growth abnormalities if the level of an environmental factor varies greatly from its optimum, or severely impacts on another factor.

Plant species have different optima and abilities to tolerate deviations from optimum levels. Furthermore, various plant processes (photosynthesis, nutrient uptake, storage of food reserves, leaf growth, etc.) may have different optima for the same environmental factor. Similarly, a particular process may have different optima for an environmental factor at various stages of plant development. Tissues or organs on the same plant may also vary in their optima for a particular environmental factor. With respect to tolerance of environmental factors, three groups of plants have been recognised on the basis of the biochemical pathways used in photosynthesis: C_3 or Calvin cycle plants, C_4 or dicarboxylic acid cycle plants and CAM or crassulacean acid cycle plants. Ca and CAM plants have lower transpiration rates than C_3 plants. They also tolerate high light intensity and high temperature better than C_3 plants and photosynthesise more efficiently. These attributes make them more able to tolerate semi-arid subtropical and tropical conditions.

Sometimes non-pathogenic diseases show characteristic symptoms and the causal factor(s) can be readily recognised by matching the symptoms with prevailing weather conditions, cultural practices or soil properties. However, symptoms of some abiotic diseases may closely resemble those of biotic diseases. In such cases it is necessary to establish that a pathogen is not associated with the disease syndrome and then relate the disease to a specific environmental factor. Restoring the relevant abiotic factor to a range more favourable for plant development should result in a return to normal growth. For example, sunflower plants showing symptoms of phosphorus deficiency resume healthy growth after the application of phosphorus ferlilisers. Similarly, in wheat, zinc deficiency can be corrected by foliar or soil applications of zinc salts.

Abiotic diseases can be caused by adverse weather conditions, adverse properties of the growing medium as well as chemical or physical injury. These factors can affect all plants, but are usually more important for cultivated plants subjected to cultural practices such as fertilisation, irrigation or chemical spraying, especially those grown in artificial environments such as greenhouses or indoors.

Adverse Weather Conditions

Various forms of severe weather such as drought, flood, high winds, frost, hail, snow and lightning may damage or kill plants or provide access sites for pathogens. Variations in 'normal' weather conditions can also affect growth resulting in reduced yield or even death of plants. For convenience, weather conditions are often discussed as independent variables such as temperature, moisture, light, composition of the atmosphere, etc. In reality, these factors interact with each other.

Temperature

Plants generally grow at temperatures ranging from about 1 °C to 40 °C, with the optimum range usually being 20-35 °C. However, C_4 plants such as many tropical grasses, have an optimum range of 31-37 °C and can grow at temperatures up to 45-55 °C. Many plants can withstand temperatures of 2-7 °C if they are protected from wind, but these temperatures can be damaging at wind speeds above 16 km/h because of both dehydrating and cooling (wind chill) effects. Perennial plants and dormant organs such as seeds and corms can tolerate more extreme temperatures than annual plants or actively growing organs.

Temperature directly affects physiological processes such as photosynthesis, respiration, membrane permeability, water and nutrient absorption, transpiration, enzyme activity and denaturation of proteins. Because the minimum, optimum and maximum temperatures for each process in an individual plant can differ (and also differ for the same process in different plants), plant reactions to temperature can be highly variable. Generally, plants growing at temperatures close to either the upper or lower end of their range will grow poorly and produce fewer and smaller flowers and fruits.

Low temperatures cause a wide range of damage to plant tissue. The severity of the damage depends on how quickly the temperature drops and also, it is thought, on how quickly tissues thaw out after freezing. Late frosts can damage young meristematic tips, entire herbaceous plants and the buds, flowers, young fruit, new twigs of fruit trees and other deciduous trees. Low temperatures may even kill the young roots of trees or cause tree bark to split, allowing canker development. Indoor or tropical plants exposed to low temperatures may turn yellow and drop their leaves or buds. Fleshy tissues, such as potato tubers, show varying degrees of low temperature damage depending on how low and for how long the temperature drops.

Low temperature damage to plant tissues results from the formation of ice crystals inside plant cells. Ice forms first on the surface of plants and spreads rapidly in the continuous film of moisture on plant cells to the intercellular spaces. Eventually ice crystals form within cells. If the growth of the ice crystals ruptures membranes or other vital cellular structures, the affected cells die. Freezing injury does not normally occur until temperatures drop below -5 to -6 °C. However, some species of bacteria in the genera Pseudomonas, Xanthomonas and Ertuinia act as nuclei for ice formation allowing ice to form on plants at temperatures between -1 and -5 °C. These bacteria grow epiphytically on plant surfaces in temperate zones around the world and one effective icenucleating bacterial cell on a leaf is sufficient to cause ice formation in that entire leaf.

Chilling injury occurs at much higher temperatures than frost damage. It is not related to the formation of ice crystals, but to the disruption of enzyme systems which causes a build-up of plant by-products which essentially 'poison' the plant. Symptoms of chilling, depending on severity, can appear immediately after chilling has occurred, or be delayed for some time. Chilling injury occurs commonly in tropical foliage plants such as Helicon as far north as Cairns in north Queensland (Latitude 17 °S). Subtropical plants are damaged at 2-8 °C while tropical plants may be injured at 12-13 °C. Minor chilling damage in plants usually results in water soaking, yellowing and browning along the edges of leaves. Older leaves show symptoms first. If the damage is severe, plants wilt and often die. Chilling injury also occurs in a wide range of fruits and vegetables if they are stored at sub-optimal temperatures after harvest. Tropical fruits may be damaged at temperatures below 10-15 °C while apples can be safely stored at 2-3 °C. Oranges stored below their optimum temperature of 5 °C may develop sunken, brown spots or pits while cucumbers stored at or below 7 °C develop surface pitting or dark-coloured water-soaked areas under the skin.

High-temperature injury results from the coagulation and denaturation of proteins, the disruption of cell membranes and the possible release of toxic products into cells. Affected cells may be killed and affected tissues desiccated. As the synthesis of normal protein slows, new proteins called 'heat shock proteins' are produced which help plants resist the effects of high temperature and other environmental stresses. The severity of high temperature damage depends on the temperature and the duration of exposure. It may take several weeks for symptoms to develop fully by which time it is difficult to relate the symptoms to the stress itself.

Plants show a variety of responses to high temperature. Seedlings may show symptoms of 'stem girdling' due to radiating heat from the soil scorching young stem tissue. If the stem above the injured zone swells, a 'heat canker' is formed. High temperatures during summer may cause many of the vegetative buds of plants such as almonds to abscise. The buds that survive either grow vigorously or remain dormant leading to very uneven growth. The flesh of ripening stone fruit darkens, especially near the stone, when temperatures exceed 39-40 °C. If high temperatures are combined with high winds, the fruit may shrivel due to combined heat and water stress.

In enclosed greenhouses, high temperatures create high humidity which significantly retards transpiration, even though the stomates remain open. Since plants normally rely on transpiration for cooling, these conditions can result in heat injury.

Humidity

The moisture content of the atmosphere is usually described by its relative humidity, the amount of moisture in the air expressed as a percentage of the amount that would be necessary to fully saturate the air at the same temperature.

Low humidity increases the evaporative demands on plants to the extent that moisture stress can occur, even when there is an ample supply of water to the roots. Normally, the evaporation of water from leaf cells in the process of transpiration cools leaves, preventing them from reaching temperatures at which damage occurs. If the supply of water becomes inadequate, the concentration of abscisic acid rises rapidly, causing the stomates to close. Although closing the stomates protects the plant from excessive water loss, it may, at the same time, reduce the degree of cooling of the plant through transpiration. The overall result is tissue damage.

Lack of moisture in the atmosphere is rarely a problem in outdoor plants. It is usually only a temporary condition and seldom causes serious damage. However, if low relative humidity is combined with strong winds, high temperatures and low soil moisture, plants may wilt temporarily or permanently, leaves may be scorched or burned and fruit shrivelled and burned due to excessive loss of water. Plants grown in greenhouses are more prone to damage by low humidity than plants grown out-of-doors. Misting or fogging is frequently used to maintain the relative humidity in the range appropriate to the plants under cultivation (70-85% is suitable for most plants). Misting or fogging also helps to reduce temperature.

When the relative humidity is high, evaporation from the plant may be suppressed, inhibiting the uptake of nutrients, particularly calcium, from the growing medium. The resulting lack of nutrients impairs cell formation and leaves plants vulnerable to a range of disorders. Excessive moisture in the atmosphere may also damage plants directly. For example, the skins of young immature cherries may split during periods of prolonged wet weather, exposing the fruit to the risk of infection by the brown rot fungus.

Atmospheric Gases

In the field, concentrations of atmospheric gases are unlikely to limit plant growth. However, in sealed glasshouses, carbon dioxide can become a limiting factor for plant growth. Adding carbon dioxide during daylight hours increases yields, improves quality and shortens cropping times. On a global scale, the amount of carbon dioxide in the atmosphere is rising and is expected to double towards the end of the 21st century. It is anticipated that this increase in the carbon dioxide content of the air will favour the growth of C_3 plants rather than C_4 plants. However, the rise in temperature (1.5-4.5 °C) that will accompany the increase in carbon dioxide concentration should encourage the growth of C_4 plants. The effects of these changes on the occurrence and severity of damage caused by pathogens are currently being investigated.

In storage, tissues in the centres of fleshy fruits and vegetables may suffer from a shortage of oxygen and a build up of carbon dioxide if the storage area is not correctly ventilated. Blackheart of potato, heart leaf injury in lettuce and brown heart of apples and pears are all attributed to the excessive accumulation of carbon dioxide in internal tissues during storage. High levels of carbon dioxide in storage may also result in external damage which usually takes the form of numerous small sunken pits on the surfaces of fruits and vegetables. While high levels of carbon dioxide may damage stored produce, properly maintained carbon dioxide levels in storage areas delay ripening and senescence, reduce the sensitivity of stored produce to ethylene-induced changes, help reduce physiological diseases and decay and control insects. This is the principle underlying the use of controlled atmosphere (CA) for the long term storage of produce such as apples and pears. Modified atmosphere (MA), a less precise form of storage than CA, is used in storage and transport of harvested fruits and vegetables.

A large number of phytotoxic chemicals, such as ethylene, nitrogen dioxide, peroxyacetyl nitrates (PANS), ozone, photochemical smog, hydrogen fluoride and sulphur dioxide, occur in the air over industrial or urban areas. All these substances, if sufficiently concentrated, can damage leaves of susceptible plants. In fact, plants sensitive to particular pollutants may be used to monitor their levels in the atmosphere. Of Australian native plants, species of Eucalyptus appear to be most susceptible to sulphur dioxide while Casuarina spp. are most resistant. Most native species tested are resistant to ozone damage although a few species of Banksia, Acacia, Melalence and Eucalyptus developed acute foliar symptoms. Plants of any type growing near industrial installations such as fertiliser plants and smelters are likely to be damaged by emission gases such as fluorides and sulphur dioxide.

Oxides of sulphur and nitrogen may be washed out of the atmosphere by rain as sulphuric and nitric acids, respectively, in the phenomenon known as acid rain. Acid rain has a pH below that of 'normal' rain (pH 5.6) and can damage plants and other forms of life over wide areas. Even if plants are not damaged severely enough to show symptoms, the pollutants may interfere with their metabolism, reducing growth, productivity and ability to reproduce.

More localised damage may occur in enclosed areas such as transport and storage areas and homes. Under these circumstances, gases produced by plants or plant products (e.g. ethylene, a plant growth regulator), by leaks in cooling systems (e.g. ammonia) or by engines (e.g. ethylene) may damage plants. For example, flowers of Geraldton Wax (Chamaelaucium uncinatum) packed in boxes for transport, frequently fall because ethylene produced by flowers infected by fungi such as Botrytis spp. or subjected to water stress after picking, stimulates flower abscission. Similarly, if fruits or vegetables that produce ethylene as they ripen are stored with those that need ethylene to ripen, considerable losses may occur as the products ripen, and possibly rot, during transport or storage. Grapes are shipped in boxes containing chemicals that liberate the fungistatic gas, sulphur dioxide. If the concentration of sulphur dioxide reaches too high a level, sunken, bleached areas sharply demarcated from unaffected tissue will appear on the fruit.

Light

Disease attributed to the effects of 'light' are often difficult to separate from the effects of other environmental factors, such as temperature. When plants grown in the shade are exposed to excessive light intensities, abnormal photochemical reactions which inactivate some enzymes and oxidise chlorophyll occur. These effects tend to be more pronounced if oxygen is freely available. Such photooxidation processes can lead to chlorosis or bleaching of plant parts and sometimes death of leaves or even the whole plant. In contrast, potato tubers exposed to light during growth or after harvest develop a green colouration due to the formation of chlorophyll in tissues that do normally contain chlorophyll. Other compounds such as the glycoalkaloid solanine may also form in exposed tissues. Solanine gives potatoes a bitter taste and is poisonous if consumed in large quantities.

The contribution of ultraviolet light to photo-oxidation processes is largely unknown. However, ultraviolet light has been implicated in sunscald of bean pods at high altitudes, in sunscald or atmospheric scorch of peanuts and in solar injury of melons. Globally, the reduction in the amount of ozone in the stratosphere will be accompanied by an increase in the amount of UV-B radiation (280-320 nm) reaching the earth. Preliminary studies indicate that increased UV-B radiation may suppress photosynthesis and, consequently, reduce plant growth.

Insufficient light inhibits chlorophyll formation and stimulates cell elongation resulting in etiolated plants with abnormally long internodes and poorly developed, pale green leaves. Etiolated plants are susceptible to lodging and invasion by parasites. They are not common out-of-doors, but may be found under enclosed conditions such as seed beds and greenhouses if light is limited.

Adverse Properties of the Growing Medium

Plants cannot function properly if the physical or chemical properties of the medium in which they are growing prevent their root systems from obtaining water, nutrients or oxygen. The components of the medium interact with each other to produce the conditions under which the plant must grow. Interactions between various properties of the growing medium may cause some plant requirements to exceed the tolerance range of the plant for that particular requirement. Accordingly, plant disease symptoms arise from a deficiency, excess or imbalance of the requirement.

SOIL-BORNE PLANT PATHOGENS

Soils contain diverse communities of microscopic organisms that are capable of damaging plants. A detrimental interaction between a soil organism and a plant is often highly specific. For example, a fungus that causes root-rot of wheat may have no effect on the roots of another plant growing in the same soil. Highly specialised interactions between soil organisms and plants can kill seedlings and even adult trees. Many organisms target younger plants but others appear as problems at later stages in the life of the plant. Other pathogens are able to cause disease in many different plant species.

The soil organisms that have the potential to be plant pathogens include fungi, bacteria, viruses, nematodes and protozoa. Some pathogens of the above ground parts of plants (leaves, stems) survive in the soil at various stages in their life cycles. Therefore, a soil phase of a plant pathogen may be important, even if the organism does not infect roots.

In spite of the potential for severe damage to be inflicted on plants by soil organisms, most plants do not display serious symptoms of disease. Disease usually occurs when conditions are particularly unfavourable, or when a soil organism is accidentally introduced where a highly susceptible plant species occurs. Intensive production in agriculture, horticulture or forestry increases the opportunities for diseases to develop compared with undisturbed natural ecosystems. Planting of similar plant species together in monoculture increases the probability of disease outbreak. In contrast, the damage caused by the fungus Phytophthora cinnamomi in many different plant species in diverse natural ecosystems in Australia demonstrates the damage that can be caused by a pathogen that infects the roots of many unrelated plants.

Some plant pathogens depend on their host plant for survival and are unable to complete their life cycle without infecting their host plant. Biotrophic organisms of this type are often difficult to grow in laboratory media.

Disease-causing microorganisms and soil animals are a natural component of the soil community. The organisms are normally present in relatively low numbers. An outbreak of disease commonly follows either an increase in the abundance of the pathogen or a change in the susceptibility of the host to the pathogen.

Examples of Major Soil-Borne Plant Diseases

Disease caused by Phytophthora Cinnamomi

Phytophthora cinnamomi causes a serious disease that threatens forests and other ecosystems especially in south-eastern and south-western Australia. Hundreds of different plant species are killed by this introduced pathogen. Vehicles are often responsible for the widespread distribution of the pathogen by disturbing and transporting infected soil. There is no simple solution to Phytophthora disease in the forest. Quarantine methods have been introduced to limit the spread of the fungus and cleaning of vehicles is mandatory.

P. cinnamomi and related fungi also cause disease of horticultural plants such as azalea, pineapple and avocado. In horticultural systems, organic mulches have been used to stimulate the community of soil organisms and reduce the negative impact of the pathogens.

Take-all Disease of Wheat

Take-all disease is caused by the fungus Gaeumannomyces graminis var. tritici. This pathogen infects the vascular tissue of wheat roots and restricts the transport of water and nutrients within the plant. Severely infected plants have stunted root systems. In addition to root rot, a severe symptom is 'white heads' which occurs if plants survive seedling damage and grow to maturity. Such plants form seed heads with poor grain development that are characteristically white. The fungus survives in the soil on decaying plant material and relies on this material as a carbon source to sustain it until it is able to infect new roots in the following wheat crop or alternate hosts.

The take-all fungus is a major root disease world-wide and is estimated to cause millions of dollars in lost wheat yield each year. The main method for control is by crop rotation with non-host plants and removal of weeds that may act as alternate hosts. There has been little success in breeding for resistance in wheat to the fungus Gaeumannomyces graminis var. tritici. Under some conditions, soils become naturally suppressive to the fungus. This has been observed in long-term monocultures of wheat. Reduced disease is likely to be due to changes in activity of other soil microorganisms.

Crown Gall

Crown gall occurs on many genera of plants and is characterised by the formation of root tumours caused by the bacterium Agrobacterium tumefaciens. The bacteria infect the root and induce plant cells to divide; a tumour-like swelling is formed that contains infected cells around its outer surface. Bacterial DNA is transferred to the host plant.

Root Knot Nematode Disease

Root knot nematodes cause disease on hundreds of plant species, especially horticultural species, in warmer climatic zones. Species of nematodes in the genus Meloidogone induce the formation of numerous galls throughout the root system. The damaged roots also have malfunctioning root tips which reduce root growth, resulting in considerable yield losses.

Identification of Plant Disease

The fundamental approach to identifying the organism or organisms responsible for a disease is to follow the protocol known as Koch's Postulates:

Step 1. Visual inspection of the plant to identify signs of the presence of a pathogen and symptoms on the host (observed without the aid of a microscope).

Step 2. Observation of the diseased portion of the plant using a microscope.

Step 3. Isolation and purification of the pathogen (growth of the organism on artificial media).

Step 4. Inoculation of plants of the same type with organisms isolated from the diseased plant to observe symptoms of disease, for comparison with the original observations in Step 1.

The purpose of using Koch's Postulates is to identify the organism responsible for disease in a rigorous manner. This is the first step in working out how to overcome the problem. Not all organisms can be isolated from plants and grown in culture and for many diseases it is difficult to identify the responsible organism.

Once a potential pathogen has been identified, a variety of procedures can be used to determine the conditions under which the disease is likely to be most severe.

Dispersal of Plant Disease

Disease occurs under the following conditions:

i. If there are enough infective hyphae, spores or other fungal structures such as sclerotia of a pathogen present in the soil,

ii. If there is a suitable environment for growth of the organisms, and

iii. If the plant is susceptible to infection.

If the organism is present in low numbers, it may have little impact on the plant. If soil conditions are suitable for the multiplication of the pathogen, there will be a greater chance of disease if other factors are also appropriate for disease development. In general, the more organisms present, the more severe the disease, but plant and soil conditions are also important.

The hyphae of some fungal pathogens can grow long distances through the soil. For example, hyphae of the fungus Armillaria extend many metres horizontally in soil and simultaneously infect the roots of several plants. Some fungi, for example species of Phytophthora, form narrow and short hyphae that do not spread the fungus through the soil. Often these fungi are disseminated in water or in soil during erosion and by other mechanisms including transport of soil on vehicles or footwear.

Spores of fungi that are pathogens of leaves and stems may have a soil phase during which spores remain dormant. Once the dormancy period ceases, spores survive in the soil until conditions are appropriate for their germination and infection of a new plant. Growth of some pathogens is stimulated by the presence of roots which release exudates or emit other signals.

Spores of some soil-borne fungi are dispersed by wind. Dispersal of bacterial plant pathogens is common when soil is transported from one place to another. Some disease-causing organisms are dispersed on plant debris. Pleiochaeta setosa spores are dispersed in soil via dead organic material from the previous crop.

Pathogen Entry via Roots

Pathogenic invasion of plant roots involves several steps. The first of these is recognition between the pathogen and the plant, which involves complex molecular mechanisms. Recognition occurs in the diffuse interface between the root and the soil created by root exudates and sloughed off root cells. This interface between soil and root may be a very suitable habitat for growth of the pathogen before it enters the root.

This initial interaction can also be the stimulus for the fungus to penetrate the plant cell wall. Some organisms produce enzymes that dissolve cell wall layers, allowing the hyphae to enter. Pathogens can also enter roots at points damaged by soil animals or other disturbances. A plant may react to the presence of a pathogen by producing chemical or physical barriers that prevent entry or confine potential pathogens to a small part of the root.

Once inside the root, pathogens occupy specific cell types. Some fungi enter the cells that conduct nutrients, carbon and water throughout the plant. If these cells are damaged, important processes that control the life of the plant are disrupted. Other fungi predominantly colonise the cortical root cells that are important in nutrient uptake from the soil.

Root Reaction to Pathogens

Plants differ in their susceptibility to plant pathogens. Even cultivars of one plant species can be more or less resistant to a pathogen. Similarly, species or strains of pathogens differ in their capacity to infect a plant. Differences in the degree of disease can therefore be related to: (i) the susceptibility of the plant: (ii) the virulence of the pathogen; (iii) the stage of development of the plant; (iv) the abundance of the pathogen; and (v) the environmental conditions.

Plant Nutrition Affects on Disease

The nutritional status of a plant influences many of its physiological processes and this can alter the response that plants have to the presence of pathogens. Changes in the concentration of nutrients such as phosphorus in a plant can alter root and shoot growth as well as the movement of carbohydrate and other molecules within the plant. These changes have the potential to influence the development of a disease.

Fertilisers can also alter plant susceptibility to disease. in the absence of manganese fertiliser, barley grew less well if nematodes were present in the soil than when manganese was added at sufficient levels to meet the requirements for growth of the plant. This occurred even when the numbers of nematodes in both treatments were similar.

Biological Control of Plant Disease

The control of pathogens and prevention of plant disease is a natural soil biological process. Indeed, in most situations, plant disease is not strongly evident even when potentially pathogenic fungi are present in a soil. This suppressive characteristic of soil is common in natural ecosystems, but becomes less apparent with high levels of disturbance and the introduction of plant monocultures. Disturbance creates soil conditions and a high density of susceptible roots that encourages the multiplication of pathogens. Once potentially damaging organisms become present in high numbers in a soil, they may be difficult to eradicate. Management practices are required that create conditions in the soil that are not favourable to pathogens so that their growth is limitedand therefore, disease it restricted.

There are many demonstrations of the potential for soil organisms to reduce disease. when apple trees were replanted in a soil previously prone to apple replant disease, increased tree growth occurred when various strains of bacteria were added to the soil, indicating a beneficial effect. Similar effects were observed at two sites, but because different quantities of the bacteria were added in the inoculum it is difficult to compare the effectiveness of each organism.

In some environments, take-all disease of wheat has declined naturally. These areas of natural suppression of disease have been investigated extensively to determine how the pathogen is suppressed. Specific bacteria that are capable of limiting the growth of the fungus, including Pseudomonas species, have been isolated from disease suppressive soils. But it is still unclear whether these bacteria are responsible for the decline in take-all disease.

NUTRIENT DEFICIENCY

Growing plants act as integrators of all growth factors and are the products in which the grower is interested. Therefore, careful inspection of the growing plant can help identify a specific nutrient stress. if a plant is lacking in a particular nutrient, characteristic symptoms may appear. Deficiency of a nutrient does not directly produce symptoms. Rather, the normal plant processes are thrown out of balance, with an accumulation of certain intermediate organic compounds and a shortage of others. This leads to the abnormal conditions recognized as symptoms. Visual evaluation of nutrient stress should be used only as a supplement to other diagnostic techniques (i.e. soil and plant analysis). Nutrient deficiency symptoms may be classified as follows:

- Complete crop failure at the seedling stage.
- Severe stunting of plants.
- Specific leaf symptoms appearing at varying times during the season.
- Internal abnormalities such as clogged conductive tissues.
- Delayed or abnormal maturity.
- Obvious yield differences, with or without leaf symptoms.
- Poor quality of crops, including differences in protein, oil, or starch content, and storage quality.
- Yield differences detected only by careful experimental work.

Each symptom must be related to some function of the nutrient in the plant. A given nutrient may have several functions, which makes it difficult to explain the physiological reason for a particular deficiency symptom. For example, when N is deficient, the leaves of most plants become pale green or light yellow. When the quantity of N is limiting, chlorophyll production is reduced, and the yellow pigments, carotene and xanthophylls are shown through a number of nutrient deficiencies produced such as pale green or yellow leaves, and the deficiency must be further related to a particular leaf pattern or location on the plant.

Apparent visual deficiency symptoms can be caused by many factors other than a specific nutrient stress. Precautions in interpreting nutrient deficiency symptoms include the following:

- The visual symptom may be caused by more than one nutrient. For example, N-deficiency symptoms may be identified, although S may also be deficient and its symptoms may not be readily apparent. B deficiency is accompanied by a red coloration of the leaves near the growing point when the plant is well supplied with K. On the other hand, when the K content is low, yellowing of alfalfa leaves occurs.
- Deficiencies are actually relative, and a deficiency of one nutrient may be related to an excessive quantity of another. For example, Mn deficiency may be induced by adding large quantities of Fe, provided that soil Mn is marginally deficient. Also, at a low level of P supply, the plant may not require as much N compared to normal or adequate P. In other words, once the first limiting factor is eliminated, the second limiting factor will appear (Liebig's law of the minimum).
- It is often difficult to distinguish among the deficiency symptoms in the field, as disease or insect damage can resemble certain micronutrients deficiencies. For example, leaf hopper damage can be confused with deficiency in alfalfa.
- A visual symptom may be caused by more than one factor. For example, sugars in corn combine with flavones to form anthocyanins (purple, red, and yellow pigments) and their accumulation may be caused by an insufficient supply of P, low soil temperature, insect damage to the roots, or N deficiency.

Nutrient deficiency symptoms appear only after the nutrient supply is so low that the plants can no longer function properly. In such cases, it would have been profitable to have applied fertilizer long before the symptoms appeared. If the symptoms are observed early, it might be corrected during the growing season. Since the objective is to get the limiting nutrient into the plant as quickly as possible, with some nutrients and under some conditions this may be accomplished with foliar applications or side dressings. Usually the yield is reduced below the quantity that would have been obtained if adequate nutrients had been available at the beginning. However, if the problem is properly diagnosed, the deficiency can be corrected the following year.

Hidden Hunger

Hidden hunger refers to a situation in which a crop needs more of a given nutrient yet has shown no deficiency symptoms. The nutrient content is above the deficiency symptom zone but still considerably needed for optimum crop production. With most nutrients on most crops, significant responses can be obtained even though no recognizable symptoms have appeared.

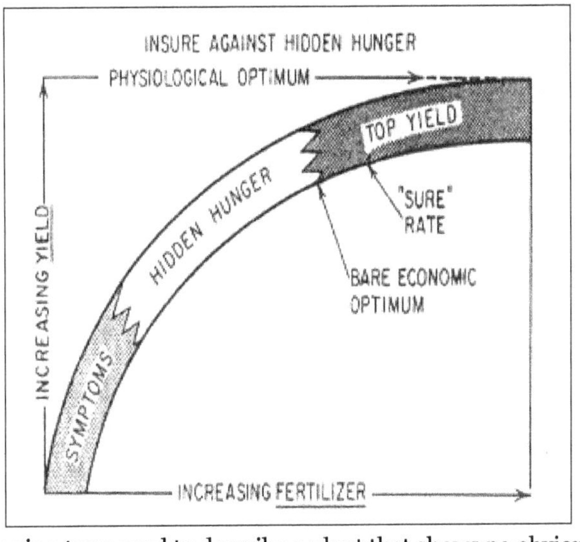

Hidden hunger is a term used to describe a plant that shows no obvious symptoms, yet the nutrient content is not sufficient to give the top profitable yield.

The question, then, is how best to eliminate hidden hunger. Testing of plants and soils is helpful for planning or modifying plant-nutrient programs to avoid this problem in subsequent crops. In both approaches, careful consideration must be given to past management practices.

Generalized Visual Leaf and Plant Nutrient Element Deficiency and Excess Symptoms:

Nitrogen (N)

- Deficiency: Light green leaf and plant color with the older leaves turning yellow, leaves that will eventually turn brown and die. Plant growth is slow, plants will be stunted, and will mature early.
- Excess: Plants will be dark green in color and new growth will be succulent; susceptible if subjected to disease and insect infestation; and subjected to drought stress, plants will easily lodge. Blossom abortion and lack of fruit set will occur.

- Ammonium toxicity: Plants fertilized with ammonium-nitrogen (NH_4- N) may exhibit ammonium-toxicity symptoms, with carbohydrate depletion and reduced plant growth. Lesions may occur on plant stems, there may be a downward cupping of the leaves, and a decay of the conductive tissue at the base of the stem with wilting of the plants under moisture stress. Blossom-end rot of fruit will occur and Mg deficiency symptoms may also occur.

Phosphorus (P)

- Deficiency: Plant growth will be slow and stunted, and the older leaves will have a purple coloration, particularly on the underside.

- Excess: Phosphorus excess will not have a direct effect on the plant but may show visual deficiencies of Zn, Fe, and Mn. High P may also interfere with the normal Ca nutrition, with typical Ca deficiency symptoms occurring.

Potassium (K)

- Deficiency: On the older leaves, the edges will look burned, a symptom known as scorch. Plants will easily lodge and be sensitive to disease infestation. Fruit and seed production will be impaired and of poor quality.

- Excess: Plants will exhibit typical Mg, and possibly Ca deficiency symptoms due to a cation imbalance.

Calcium (Ca)

- Deficiency: The growing tips of roots and leaves will turn brown and die. The edges of the leaves will look ragged as the edges of emerging leaves stick together. Fruit quality will be affected with the occurrence of blossom-end rot on fruits.

- Excess: Plants may exhibit typical Mg deficiency symptoms, and when in high excess, K deficiency may also occur.

Magnesium (Mg)

- Deficiency: Older leaves will be yellow in color with interveinal chlorosis (yellowing between the veins) symptoms. Plant growth will be slow and some plants may be easily infested by disease.

- Excess: Results in a cation imbalance showing signs of either a Ca or K deficiency.

Sulfur (S)

- Deficiency: A general overall light green color of the entire plant with the older leaves being light green to yellow in color as the deficiency intensifies.

- Excess: A premature senescence of leaves may occur.

Boron (B)

- Deficiency: Abnormal development of the growing points (meristematic tissue) with the apical growing points eventually becoming stunted and dying. Rowers and fruits will abort. For some grain and fruit crops, yield and quality is significantly reduced.
- Excess: Leaf tips and margins will turn brown and die.

Chlorine (Cl)

- Deficiency: Younger leaves will be chlorotic and plants will easily wilt. For wheat, a plant disease will infest the plant when Cl is deficient.
- Excess: Premature yellowing of the lower leaves with burning of the leaf margins and tips. Leaf abscission will occur and plants will easily wilt.

Manganese (Mn)

- Deficiency: Interveinal chlorosis of young leaves while the leaves and plants remain generally green in color. When severe, the plants will be stunted.
- Excess: Older leaves will show brown spots surrounded by a chlorotic zone and circle.

Molybdenum (Mo)

- Deficiency: Symptoms will frequently appear similar to N deficiency. Older and middle leaves become chlorotic first, and In some instances, leaf margins are rolled and growth and flower formation are restricted.
- Excess: Not of common occurrence.

Zinc (Zn)

- Deficiency: Upper leaves will show interveinal chlorosis with an eventual whiting of the affected leaves. Leaves may be small and distorted with a rosette form.
- Excess: An Fe deficiency will develop.

TEMPERATURE EXTREMES

Rate of plant growth and development is dependent upon the temperature surrounding the plant and each species has a specific temperature range represented by a minimum, maximum, and optimum. These values were summarized by Hatfield et al. for a number of different species typical of grain and fruit production. The expected changes in temperature over the next 30-50 years are predicted to be in the range of 2-3 °C. Heat waves or extreme temperature events are projected to become more intense, more frequent, and last longer than what is being currently been observed in recent years. Extreme temperature events may have short-term durations of a few days with temperature increases of over 5 °C above the normal temperatures. Extreme events occurring during the summer period would have the most dramatic impact on plant productivity; however, there has been little research conducted to document these effects as found by Kumudini et al. A recent review by Barlow et al. on the effect of temperature extremes, frost and heat, in wheat revealed that frost caused sterility and abortion of formed grains while excessive heat caused reduction in grain number and reduced duration of the grain-filling period. Analysis by Meehl et al. revealed that daily minimum temperatures will increase more rapidly than daily maximum temperatures leading to the increase in the daily mean temperatures and a greater likelihood of extreme events and these changes could have detrimental effects on grain yield. If these changes in temperature are expected to occur over the next 30 years then understanding the potential impacts on plant growth and development will help develop adaptation strategies to offset these impacts.

Temperature Responses

Responses to temperature differ among crop species through-out their life cycle and are primarily the phenological responses, i.e. stages of plant development. For each species, a defined range of maximum and minimum temperatures form the boundaries of observable growth. Vegetative development (node and leaf appearance rate) increases as temperatures rise to the species optimum level. For most plant species, vegetative development usually has a higher optimum temperature than for reproductive development. Cardinal temperature values for selected annual (non-perennial) crops are given in Hatfield et al. for different species. If we depict the range of temperatures in the following diagram then the definition of extreme temperatures affecting plant response will be species dependent. For example, an extreme event for maize (Zea mays L.) will be warmer than for a cool season vegetable (broccoli) where the maximum temperature for growth is 25 °C compared to 38 °C.

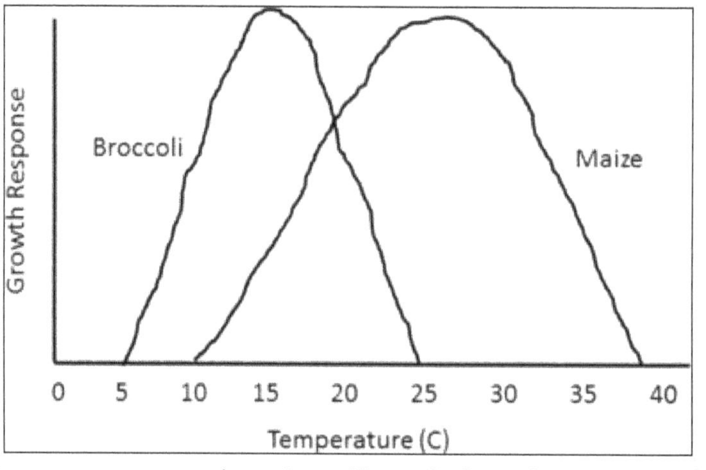

Temperature response for maize and broccoli plants showing the lower, upper and optimum temperature limits for the vegetative growth phase.

In understanding extreme events and their impact on plants we will have to consider the plant temperature response relative to the meteorological temperature.

Faster development of non-perennial crops results in a shorter life cycle resulting in smaller plants, shorter reproductive duration, and lower yield potential. Temperatures which would be considered extreme and fall below or above specific thresholds at critical times during development can significantly impact productivity. Photoperiod sensitive crops, e.g. soybean, would also interact with temperature causing a disruption in phonological development. In general, extreme high temperatures during the reproductive stage will affect pollen viability, fertilization, and grain or fruit formation. Chronic exposures to extreme temperatures during the pollination stage of initial grain or fruit set will reduce yield potential. However, acute exposure to extreme events may be most detrimental during the reproductive stages of development.

The impacts of climate change are most evident in crop productivity because this parameter represents the component of greatest concern to producers, as well as consumers. Changes in the length of the growth cycle are of little consequence as long as the crop yield remains relatively consistent. Yield responses to temperature vary among species based on the crop's cardinal temperature requirements. Warming temperatures associated with climate change will affect plant growth and development along with crop yield.

Temperature Extremes in Climate

One of the more susceptible phenological stages to high temperatures is the pollination stage. Maize pollen viability decreases with exposure to temperatures above 35 °C. The effect of temperature is enhanced under high vapor pressure deficits be-cause pollen viability (prior to silk reception) is a function of pollen moisture content which is strongly dependent on vapour pressure deficit. During the endosperm division phase, as temperatures increased to 35 °C from 30 °C the potential kernel growth rate was reduced along with final kernel size, even after the plants were returned to 30 °C. Exposure to temperatures above 30 °C damaged cell division and amyloplast replication in maize kernels which reduced the size of the grain sink and ultimately yield. Rice shows a similar temperature response to maize because pollen viability and production declines as daytime maximum temperature (T_{max}) exceeds 33 °C and ceases when T_{max} exceeds 40 °C. Current cultivars of rice flower near mid-day which makes T_{max} a good indicator of heat-stress on spikelet sterility. These exposure times occur quickly after anthesis and exposure to temperatures above 33 °C within 1-3 h after anthesis (dehiscence of the anther, shedding of pollen, germination of pollen grains on stigma, and elongation of pollen tubes) cause negative impacts on reproduction.

Current observations in rice reveal that anthesis occurs between about 9 to 11 am in rice and exposure to high temperatures may already be occurring and will increase in the future. There is emerging evidence that differences exist among rice cultivars for flowering times during the day. Given the negative impacts of high temperatures on pollen viability, recent observations from Shah et al. suggest flowering at cooler times of the day would be beneficial to rice grown in warm environments. They proposed that variation in flowering times during the day would be a valuable phenotypic marker for high-temperature tolerance. As daytime temperatures increased from 30 to 35 °C, seed set on male-sterile, female fertile soybean (Glycine max (L.)Merr.) plants decreased. This confirms earlier observations on partially male-sterile soybean in which complete sterility was observed when the daytime temperatures exceeded 35 °C regardless of the night temperatures and concluded that daytime temperatures were the primary factor affecting pod set. Crop sensitivity to temperature extremes depends upon the length of anthesis. Maize, for example, has a highly compressed phase of anthesis for3-5 days, while rice, sorghum (Sorghum bicolor L. Moench.) and other small grains may extend anthesis over a period of a week or more. Inn soybean, peanut (Arachis hypogaea L.), and cotton (Gossypium hirsutum L.) anthesis occurs over several weeks and avoid a single occurrence of an extreme event affecting all of the pollening flowers. For peanut (and potentially other legumes) the sensitivity to elevated temperature for a given flower, extends from 6 days prior to opening (pollen cell division and formation) up through the day of anthesis. Therefore, several days of elevated temperature may affect fertility of flowers in their formative 6-day phase or anthesis. found differences in the threshold temperature for grain sorghum among genotypes and differences in the percentage of seed set in response to high temperatures. Pollination processes in other cereals, maize and sorghum, may have a similar sensitivity to elevated daytime temperature as rice. Rice and sorghum have exhibited similar sensitivities of grain yield, seed harvest index, pollen viability, and success in grain formation in which pollen viability and percent fertility is first reduced at instantaneous hourly air temperature above 33 °C and reaches zero at 40 °C. Diurnal max/min day/night temperatures of 40/30 °C (35 °C mean) cause zero yield for those two species with the same expected response for maize.

Annual Crops

Projected air temperature increases throughout the remainder of the 21st century suggests that grain yields will continue to decrease for the major crops because of the increase temperature stress on all major grain crops. Beyond a certain point, higher air temperatures adversely affect plant growth, pollination, and reproductive processes. However, as air temperatures rise be-yond the optimum, instead of falling at a rate commensurate with the temperature increase, crop yield losses accelerate. For example, an analysis indicated yield growth for corn, soybean, and cotton would gradually increase with temperatures up to 29 °C to 32 °C and then sharply decrease with temperature increases beyond this threshold.

Increases of temperature may cause yield declines between 2.5% and 10% across a number of agronomic species through out the 21st century. Other evaluations of temperature on crop yield have produced varying outcomes. Lobell et al. showed estimates of yield decline between 3.8% and 5% and Schlenker and Roberts used a statistical approach to estimate wheat, corn, and cotton yield declines of 36% to 40%under a low CO_2 emissions scenario, and between 63% to 70% for high CO_2emission scenarios. These estimates of yield loss did not consider the positive effects of rising atmospheric CO_2on crop growth, variation among crop genetics, impact of biotic stresses on crop growth and yield, or the use of adaptive management strategies, e.g. fertilizers, rotations, tillage, or irrigation. These analyses assumed that air temperature increased without regard to the potential negative effects of temperature extremes.

The current evaluations of the impact of changing temperature have focused on the effect of average air temperature changes; however, increases in minimum air temperature may be more significant in their effect on growth and phenology. Minimum air temperatures are more likely to increase under climate change. While maximum temperatures are affected by local conditions, especially soil water content and evaporative heat loss as soil water evaporates, minimum air temperatures are affected by mesoscale changes in atmospheric water vapor content. Hence, in areas where changing climate is expected to cause increased rainfall or where irrigation is predominant, large increases of maximum temperatures are less likely to occur than in regions prone to drought. Minimum air temperatures affect night time plant respiration rates and can potentially reduce biomass accumulation and crop yield. Welch et al. found higher minimum temperatures reduced grain yield in rice, while higher maximum temperature raised yields; because the maximum temperature seldom reached the critical optimum temperature for rice. However, under the scenario of future temperatures increases, they found maximum temperatures could decrease yields if they are near the upper threshold limit.

Similar responses have been found in annual specialty crops in which temperature is the major environmental factor affectting production with specific stresses, such as periods of hot days, overall growing season climate, minimum and maximum daily temperatures, and timing of stress in relationship to develop-mental stages having the greatest effect. When plants are subjected to mild heat stress (1 °C to 4 °C above optimal growth temperature), there was moderately reduced yield. In these plants, there was an increased sensitivity heat stress 7 to 15 days before anthesis, coincident with pollen development. Subjecting plants to a more intense heat stress (generally greater than 4 °C above optimum) resulted in severe yield loss extending to complete crop failure. Tomatoes under heat stress fail to produce viable pollen while their leaves remain active. The non-viable pollen does not pollinate flowers causing failure in fruit set. If the same stressed plants are cooled to normal temperatures for 10 days before pollination, and then returned to high heat, they are able to develop fruit. There are some heat tolerant tomatoes which perform better than others related to their ability to successful pollinate even under adverse conditions.

Perennial crops have a more complex relationship to temperature than annual crops. Many perennial crops have a chilling requirement in which plants must be exposed to a number of hours below some threshold temperature before flowering can occur. For example, chilling hours for apple range from 400 to 2900 h (5-7 °C base), while cherry trees (Prunus avium) require 900 to 1500 h with the same base temperature. Grapes have a lower chilling threshold that other perennial plants with some varieties being as low at 90 h. Increasing winter temperatures may prevent chilling hours from being obtained and projections of warmer winters in California revealed that by mid-21st century, plants requiring more than 800 h may not be exposed to sufficient cooling except in very small areas of the central Valley. Climate change will impact the chilling requirements for fruits and nut trees. Hatfield et al. showed that under a warming climate, adequate chilling hours for perennial crops for fruit development may not be met. Innovative adaptation strategies will be required to overcome this effect because of the long time requirements for genetic selection and fruit production once perennial crops are established.

Perennial plants are also susceptible to exposure to increasing temperatures similar to annual plants. These responses and the magnitude of the effects are dependent upon individual species. Exposure to high temperatures, 422 °C, for apples during re-production increases the fruit size and soluble solids but decreases firmness as a quality parameter. In cherries, increasing the temperature 3 °C above the 15 °C optimum mean temperature decreases fruit set. Optimum temperature range in citrus is 22-27 °C and temperatures greater than 30 °C increased fruit drop. During fruit development when the temperatures exceed the optimum range of 13-27 °C with temperatures over 33 °C there is a reduction in Brix (sugar content), acid content, and fruit size in citrus. Temperature stresses on annual and perennial crops have an impact on all phases of plant growth and development.

Exposure of plants to extreme temperatures will limit the ability of the plant to produce fruit due to disruption of the pollination process. The magnitude of this impact varies among species; however, there is a consistent negative impact on plants. One aspect of high temperature extremes often overlooked is the effect of extreme events on the atmospheric water vapor demand. If we plot the saturation vapor pressure (e*) relative to air temperature we see an exponential increase of e*with temperature. The following equation to represents the energy balance of a leaf:

$$S_t(1-a)+L_d-\varepsilon\sigma T_l^4=\frac{pC_p(T_l-T_a)}{r_a}+\frac{pC_p(e^*-e_a)}{r(r_s+r_a)}$$

Here S_t is the incoming solar radiation (W m^{-2}), α_l is the albedo of the leaf or canopy, L_d is the incoming long-wave radiation (W m^{-2}), ε is the emissivity of the leaf or canopy, σ the Stefan-Boltzmann constant, T_l the leaf or canopy temperature, r_a is the aerodynamic conductance (m s^{-1}), and r_s the canopy conductance (m s^{-1}), then we see how changing e* would affect the energy balance.

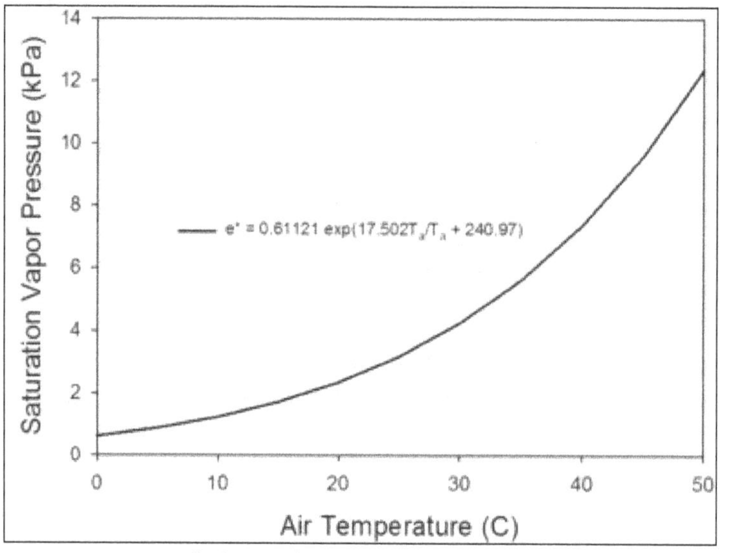

Saturation vapor pressure relative to air temperature showing the exponential increase in saturation vapor pressure with temperature.

An increasing water vapor demand will cause more water to be transpired by the leaf until the water supply becomes limited and the stomatal conductance will decrease leading to a higher leaf temperatures and a reduction in photosynthesis. If the plant is exposed to extreme temperatures, water stress could occur quickly because the plant lacks sufficient capacity to extract water from the soil profile to meet the increased atmospheric demand.

TOXIC CHEMICALS

Many complex chemicals are routinely applied to plants to prevent attack by insects, mites, and pathogens; to kill weeds; or to control growth. Serious damage may result when fertilizers, herbicides, fumigants, growth regulators, antidesiccants, insecticides, miticides, fungicides, nematicides, and surfactants (substances with enhanced wetting, dispersing, or cleansing properties, such as detergents) are applied at excessive rates or under hot, cold, or slow-drying conditions.

Some pollutants are the direct products of industry and fuel combustion, while others are the result of photochemical reactions between products of combustion and naturally occurring atmospheric compounds. The major pollutants toxic to plants are sulfur dioxide, fluorine, ozone, and peroxyacetyl nitrate.

Sulfur dioxide results primarily from the burning of large amounts of soft coal and high-sulfur oil. It is toxic to a wide range of plants at concentrations as low as 0.25 part per million (ppm) of air (i.e. on a volume basis, one part per million represents one volume of pure gaseous toxic substance mixed in one million volumes of air) for 8 to 24 hours. Gaseous and particulate fluorides are more toxic to sensitive plants than is sulfur dioxide because they are accumulated by leaves. They are also toxic to animals that feed on such foliage. Fluorine injury is common near metal-ore smelters, refineries, and industries making fertilizers, ceramics, aluminum, glass, and bricks.

Ozone and peroxyacetyl nitrate injury (also called oxidant injury) are more prevalent in and near cities with heavy traffic problems. Exhaust gases from internal combustion engines contain large amounts of hydrocarbons (substances that principally contain carbon and hydrogen molecules—gasoline, for example). Smaller amounts of unconsumed hydrocarbons are formed by combustion of fossil fuels (e.g. coal, oil, natural gas) and refuse burning. Ozone, peroxyacetyl nitrate, and other oxidizing chemicals (smog) are formed when sunlight reacts with nitrogen oxides and hydrocarbons. This pollutant complex is damaging to susceptible plants many kilometres from its source. Ozone and peroxyacetyl nitrate are capable of causing injury if present at levels of 0.01 to 0.05 part per million for several hours.

Phytotoxicity

Phytotoxicity is a toxic effect by a compound on plant growth. Such damage may be caused by a wide variety of compounds, including trace metals, salinity, pesticides, phytotoxins or allelochemicals.

Substances with Phytotoxic Potential

Inorganic Compounds

High concentrations of mineral salts in solution within the growing medium can have phytotoxic effects. Sources of excessive mineral salts include infiltration of seawater and excessive application of fertilizers. For example, urea is used in agriculture as a nitrogenous fertilizer, but if too much is applied, phytotoxic effects can result, either by urea toxicity or by the "ammonia produced through hydrolysis of urea by soil urease". Acid soils may contain high concentrations of aluminium (as Al^{3+}) and manganese (as Mn^{2+}) which can be phytotoxic.

Herbicides

Fertilizers and herbicides are the most widely used chemicals in farmlands and have been instrumental in the tremendous increase in crop productivity since World War II. However, there has also been growing concerns about declining plant species richness, abundance and diversity both within crop fields and in adjacent habitats including field margins, hedgerows, ditches, as well as small woodlots and wetlands. Many plant species associated with agro ecosystems have become rare to the extent that they are registered in the Red Data Books (International Union for Conservation of Nature) of several countries, including several arable species considered agricultural weeds. Failure to adequately assess and properly regulate herbicide effects can have important ecological implications for plant survival, seed production, long-term seed bank replenishment and eventual species composition of not only primary producers, but also species at other trophic levels.

Herbicides used in agriculture for weed control in major crops are primarily sprayed in May or June In most European countries, herbicides are sprayed several times in any given year depending on the crops. In Denmark, spring sown crops are usually sprayed with herbicides in April and May while autumn sown crops are sprayed in September and October. In the Netherlands, an average of 5.7 herbicides are sprayed on food crops (between three and nine depending on the crops) and 10.3 (between six and 15) herbicides per year are applied on field cultivated flower crops. Though it has not yet been quantified, it is likely that herbicides will reach weeds and non-target plants at all phenological stages depending on the application time.

When plants are sprayed in crop fields and sub lethal doses of herbicides reach non-target plant species in adjacent habitats through drift, runoff and/or volatilisation, resultant effects on sensitive species can be observed in any of four ways: a) Plants at the seedling stage during spray will have their vegetative parts affected, b) the same plants could express the effect through negative impacts on seed production at later stages, c) plants at the reproductive phase during spray have their seed production impacted or d) the vegetative parts of the F1 generation are affected. Therefore, it appears that seedlings and plant species at late vegetative and reproductive stages may be affected differently, and this is most likely influenced in turn by the type of herbicide applied.

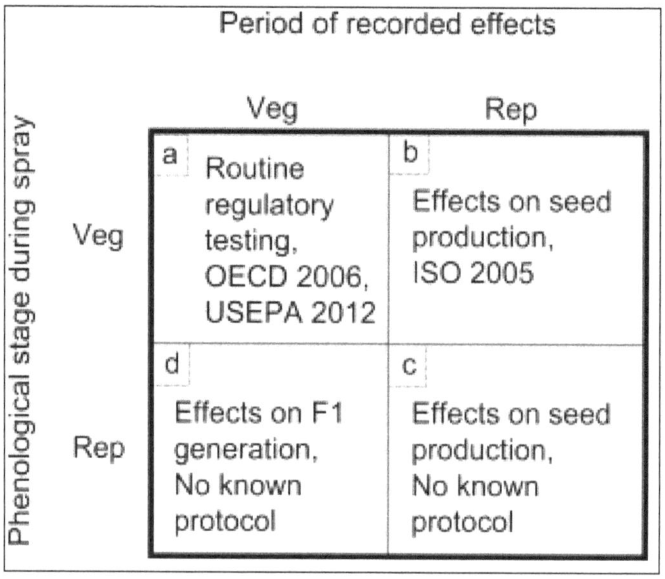

Representation of phenological stage at spray (or testing) time and stage of recorded herbicide effects. Veg = Vegetative, Rep = Reproductive period. Letters within quadrants are used for reference purposes in the text. OECD (Organisation for Economic Co-operation and Development) 2006, USEPA (United States Environmental Protection Agency) 2012 and ISO (International Organization for Standardization) 2005 refer to standard guidelines for plant toxicity testing.

For regulatory purposes, greenhouse tests utilizing species growing singly in pots or in monoculture are required to assess the potential undesirable effects of herbicides on non-target, wild plants found within the vicinity of croplands. These tests are performed on emerging seedlings or on plants at the 2–6 leaf stage (usually using crops as surrogates for wild species) with effects recorded 14–28 days after the spray event. Several greenhouse studies have effectively shown that the seedling stage was more sensitive to herbicides than later growth stages, at least for some species. However, other studies have shown that some species that have reached the reproductive stage following exposure exhibited negative herbicidal effects at doses below those observed for the seedling or vegetative stages. In some cases, reproductive endpoints may be more appropriate to assess than aboveground vegetative biomass, for instance when plants are exposed at later developmental stages when growth has ceased. The ISO protocol was developed to examine both the inhibition of growth and the reproductive capacity of plants following soil contamination (not specific to pesticides) under controlled conditions using two test species: a rapid-cycling variant of turnip rape (Brassica rapa CrGC syn. Rbr) and oat (Avena sativa L.). Though this test assesses both the vegetative and reproductive effects of contaminants on plants, it is not usually conducted for pesticide registration. There is no known protocol to determine effects when plants are sprayed at maturity.

Fungicide

The use of fungicides in agriculture for fungi diseases control has become crucial. Fungicide research has produced a diverse range of products with novel modes of action. However, the extensive use of these compounds in the agriculture system raises public concern because of the harmful potential of such substances in the environment and human health. Moreover, the phytotoxic effects of some fungicides are already recognized but little is known about the impact of these compounds on the photosynthetic apparatus.

Fungicides are chemical compounds or biological organisms that destroy or inhibit the growth of fungi or fungal spores. The use of fungicides for an effective control of plant diseases has become crucial in the last decades in the agriculture system since it is estimated that fungal infections cause yield reductions of almost 20% of crops worldwide. Due to their relatively low cost, ease of use, and effectiveness, fungicides became the primary means of fungi control. However, the extensive use of these compounds to control fungal disease in plants raises the appearance of new strains of pathogens that have become resistant to the available commercial products.

Fungicide toxicity is not always restricted to the target pest organism, having also been demonstrated in mammals including humans. The extensive use of fungicides in plant protection against fungal disease generates long-term residues in food and in the environment. In the annual EU report, EFSA (European Food Safety Authority), where vegetables and fruits of 27 countries were surveyed for pesticides contamination, the results highlighted that dithiocarbamates are among the most common residual contaminants. Thus, the abusive use of such compounds in agriculture has mobilized public concern because of the harmful potential of such substances in the environment and in the food chain representing a risk for human health.

Most of the work dealing with the impact of fungicides in agriculture is focused on their efficiency against fungal pathogens or their residues in crops. Several reports appoint that some fungicides may enhance plant defences through phytoalexin synthesis and cell wall lignification or stimulate enzymes involved in the synthesis of phenolic compounds. Others describe the putative protective role of fungicides for crops against various types of stress. Triazoles protect Hordeum vulgare and Arachis hypogaea against ozone exposure or salt stress by stimulating antioxidative enzymes. Moreover, azoxystrobin and epoxiconazole fungicides induced a delay of senescence of Triticum aestivum mainly due to an enhancement of the antioxidative potential protecting the plants from harmful active oxygen species. Researchers have described an induction of the synthesis of photosynthetic pigments and proteins in fungicide-exposed plants. However, few studies have addressed the question of whether these products alter or inhibit physiological and metabolic activities in the plant, and the negative effects of some fungicides on photosynthesis, pigment content, growth, and alterations in the reproductive organs were poorly explored. The available data report modifications on the CO_2 assimilation and photosynthetic efficiency. Photosynthesis reduction strongly conditions biomass production and growth rates, which are strictly related with crop productivity and yield. Information on fungicides effects on plant physiology (photosynthesis) is crucial for the understanding of the underlying regulatory mechanisms as a precondition to judge the phytotoxicity of a compound.

Types of Fungicides

Fungicides that are used to control plant fungi can be applied before infection to protect the plant from fungi invasion. This type of fungicides have a protective action. Others can be used to eliminate or eradicate an established infection. Fungicides can be classified in two main categories: contact (nonsystemic) and systemic fungicides.

Contact Fungicides

Contact fungicides have preventive action by killing or inhibiting fungi or fungal spores before the mycelia can grow and develop within the plant tissues. However, once the infection is established, this fungicide may not have any function. Thus, this kind of fungicides can be used only as protectants.

Inorganic copper compounds such as Bordeaux mixture and copper carbonate and inorganic sulfur in the form of elemental sulfur and lime sulfur are some examples of the main contact fungicides available for plant protection. Within the organic contact fungicides, for example, dialkyldithiocarbamates, which include the fungicides thiram, ferbam, and ziram, are a group of fungicides with a high role in the worldwide control of plant diseases since they are generally more effective and less toxic than the inorganic compounds (e.g. sulfur and copper fungicides). These multisite inhibitors have several kinds of toxic action in fungal cells such as metal chelation, mixed disulfide formation, and transport of heavy metals across membranes. Dialkyldithiocarbamates inhibit a wide range of fungal enzymes, but the pyruvic dehydrogenase system is particularly sensitive to these fungicides. Another group of organic contact fungicide widely used is the ethylenebisdithiocarbamates, which include zineb, maneb, metiram, and mancozeb. The mode of action of this type of fungicide differs from that of the dialkyldithiocarbamates: they undergo transformation to ethylenediisothiocyanate, which inactivates thiol groups of enzymes and metabolites in fungal cells.

Contact fungicides are inexpensive and fungal resistance rarely occurs. Therefore, they are still widely used for plant disease control even though many newer, more potent systemic fungicides have been developed.

Systemic Fungicides

The other category of fungicides, systemic fungicides, are absorbed by the plant and carried by translocation to the site of infection. These kind of fungicides can kill the fungus after the mycelia has penetrated the parenchyma of the plant tissue, stopping the dispersal or infection within the plant. Systemic fungicides can be used as protectants, eradicants, or both and are the most recently developed and the most promising type of fungicide for the future. However, since systemic fungicides usually have a very specific site of action in the target fungus, fungi may readily develop resistance to them if they are not managed appropriately.

Systemic fungicides comprise a wide group of compounds with several modes of action. For example, the largest and most important group of systemic fungicides used to control plant fungal diseases is the dicarboximide. The mode of action of this fungicide seems to be related to the inhibition of triglyceride biosynthesis in the fungi. The dicarboximide fungicides, iprodione, procymidone, vinclozolin, chlozolinate, and metomeclan are particularly valuable for the control of plant diseases caused by species of Botrytis, Sclerotinia, Monilinia, Alternaria, Sclerotium, and Phoma.

Benzimidazoles are a group of organic fungicides with systemic action that are also extensively used in agriculture. These types of compounds control a broad range of fungi at relatively low application rates. For example, benomyl is one of the most effective and extensively used benzimidazoles in crop protection. The benzimidazoles benomyl, carbendazim, and thiabendazole and the phenylcarbamate diethofencarb specifically interfere with the formation of microtubules, which function in a variety of cellular processes, including mitosis and maintenance of cell shape. These fungicides bind specifically to protein subunits called tubulin and prevent their assembly to form microtubules.

Since their introduction in the 1960s, systemic fungicides have gradually replaced the older nonsystemic products, establishing higher levels of disease control and developing new fungicide markets. Compared with the nonsystemic, systemic fungicides are approximately twice as valuable in terms of sales.

Plant Physiological Responses to Fungicide Exposure

The widely accepted assumption that fungicide has low phytotoxicity has started to be out dated with the publication of more detailed analysis at the cell level that demonstrated several damages at the photosynthetic apparatus (e.g.). Some reports appointed that application of fungicides has consequences on plant physiology, such as grow reduction, perturbation of reproductive organ development, alteration of nitrogen, and/or carbon metabolism. This former physiological trait is fundamental for plant culture and is reflected by both photosynthetic rate and mobilization of carbohydrate reserves. Physiological studies after fungicide application on several species reported modifications of both photosynthetic activity and chlorophyll a fluorescence. Decreased CO_2 assimilation in fungicide-treated plants has been attributed to both stomatal (due to stomatal closure) and nonstomatal effects due to a disruption in the capacity of RuBisCO carboxylation, decrease of RuBisCO content, and/or reduction of the ribulose 1.5 bisphosphate regeneration.

Net CO_2 assimilation reductions accompanied by changes in stomatal conductance and intercellular CO_2 concentration were reported in Malus domestica and Cucumis sativus after fungicide application. The application of a nonsystemic fungicide, fludioxonil, in Vitis vinifera induced a decrease in net CO_2 assimilation and in the intercellular CO_2 concentration but stomatal conductance was not affected. The application of the same fungicide, fludioxonil, and a systemic fungicide, pyrimethanil, in in vitro plants and fruiting cuttings of Vitis vinifera promoted different physiological responses: in in vitro plants, both fungicides decreased net CO_2 assimilation, transpiration rate, stomatal conductance, and intercellular CO_2 concentration; in the fruiting cuttings, the fungicides did not affect CO_2 exchange neither transpiration rates.

Some reports suggested that the systemic fungicide strobilurin may improve the water status and stress management of plants under conditions of drought stress. Nason et al. showed that the application of beta-methoxyacrylate, a strobilurin fungicide, improve the water use efficiency only in well-watered Triticum aestivum and Hordeum vulgare plants. However, when these plants are under drought stress, strobilurin strongly reduces net CO_2 assimilation, intercellular CO_2 concentration, transpiration rate, and rate of stomatal conductance to water. Net CO_2 assimilation reduction seems to be related to stomatal conductance decrease. It is possible that stomata respond to strobilurin-induced changes in mesophyll photosynthesis either by sensing changes in the intercellular CO_2 concentration or by responding to the pool size of an unidentified C-fixing substrate. It is also possible that the effects of strobilurin fungicides are mediated via ABA-based chemical signaling.

The analysis of several chlorophyll a fluorescence parameters of plants treated with fungicides demonstrated that light reactions of photosynthesis are also sensible to fungicide exposure. Bader and Abdel-Basset showed, for the first time, that fungicides of the triforine type (a systemic and contact fungicide) strongly inhibit electron-transport reactions of chloroplasts. Moreover, the application of systemic fungicides, benzimidazoles and triazole, and a dithiocarbamate contact fungicide affected the effective quantum yield of PSII (Φ_{PSII}) as well as the maximal quantum efficiency of PSII (F_v/F_m). This reduction was attributed to the decrease in photochemical quenching (qP). In Glycine max, strobilurin fungicides application reduced the ratio of F_v/F_m. Strobilurin fungicides seem to block the transport of electrons between PSII and PSI by binding to the Qi site of the chloroplast cytochrome bf complex.

Since plants depend on photosynthesis to assimilate carbon for further growth and overall vigor, photosynthesis impairment has negative consequences in plant biomass production and yield. Several reports support a decrease in biomass production in fungicides-treated plants: benomyl, a systemic fungicide, reduced the growth of Gossypium hirsutum, Helianthus annuus, Cucumis sativus, Lactuca sativa, and Pinus taeda. Moreover, the application of carbendazim (systemic benzimidazole fungicide) in Nicotiana tabacum affected negatively plant biomass.

Pigment biosynthesis is appointed by Ahmed et al. to be inhibited by the systemic fungicide, benomyl. This fungicide induces a considerable reduction on the chlorophyll a, chlorophyll b, carotenoids, and the total pigments content of Helianthus annuus plants. Similarly, the treatment of Vitis vinifera with fludioxonil and Nicotiana tabacum with carbendazim also decreases the chlorophyll and carotenoid content. Mihuta-Grimm et al. and Van Iersel and Bugbee reported leaf chlorosis after benomyl application on Impatiens walleriana, Cucumis sativus, Celosia plumose, Petunia hybrid, and Lycopersicon esculentum.

Modifications of dark respiration were reported after mancozeb (contact fungicide) and flusilazol (systemic fungicide) application in Malus domestica. The increase in dark respiration can be explained by additional energy requirement, metabolic breakdown of the compound, and/or activation of the alternative, cyanide-insensitive, respiration. Curiously, the treatment with strobilurin fungicides induced different responses: while in Triticum aestivum and in Spinacia oleracea plants respiration was inhibited, in Triticum aestivum dark respiration was reduced.

The available studies in the literature have demonstrated that fungicide application impairs photosynthesis. However, the reports available are in general based on few physiological parameters using a large variety of plant species and different types of fungicide leading in some cases to controversial results that jeopardize a comprehensive knowledge of the main photosynthetic targets of fungicides. Thus, future investigation on the subject should be considered in order to produce more reliable data to identify fungicide photosynthetic targets and build a comprehensive model of the physiological response of plant exposed to fungicides.

It is expected that fungicides remain an essential tool for plant disease management and will continue to play a crucial role in optimizing yields from the world crops. Therefore, the development of new compounds with lower negative impact in plant physiology is a future challenge. This will provide benefits not only for plants yield but also for the environment and human health.

Air Pollutants

The atmosphere, the gaseous envelope surrounding the Earth contains a number of natural constituents: a mixture of gases (78% nitrogen, 21% oxygen, and also some other minor gases including 1% argon, CO_2, neon, helium, ozone, etc.), water (solid, liquid and vapour) and solid or liquid, inorganic or organic particles in suspension (aerosols).

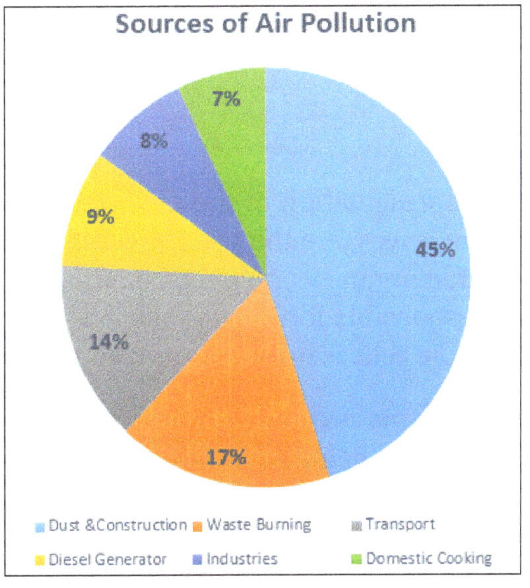

The atmosphere is polluted when the content of some of its natural constituents is higher than normal and/or when it contains new components. But air pollution is mainly referred to as air pollution when these increases lead to levels of components such that they have harmful effects on the various components of different ecosystems (plants, animals, etc.).

Depending on the type of component considered, these increases may concern very small areas as well as the entire planet. If air pollution has always existed (because of volcanism, fires, for example), it is with the advent of the industrial era that it has become a real problem for the environment and health.

Physiological Responses of Plants

While plants, due to their fixed life and wide distribution, are among the first victims of air pollution, they can also be a source of secondary pollution. During high heat, they emit volatile organic compounds (VOCs) such as terpenes, one of the precursor gases of ozone. In cities in the hot regions of the USA, it is recommended not to plant certain trees (pines, oaks, etc.) in order not to increase ozone levels. Plants also emit fine particles (pollens, spores, wax compounds, various particles) which, if they have no effect on plants, can have effects on human health (allergies).

Penetration of Pollutants into Plants

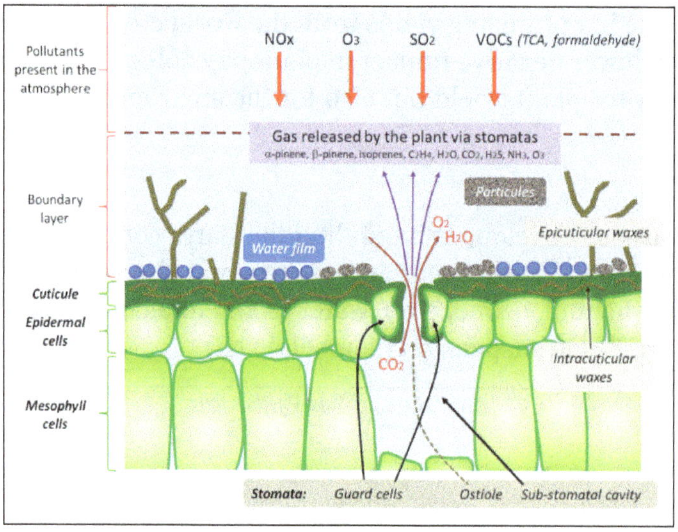

The above figure is a schematic representation of the environment of leaf surfaces. Stomata, located mainly on the surface of the leaves, are the preferred site for the plant's gas exchange with the atmosphere. Among the volatile organic compounds emitted by the plant are many terpenes -α-pinene, β-pinene, isoprene- responsible for summer air pollution, particularly in forest areas and mountain valleys. The size of the cells is in the range of 10 to 100 mm.

The penetration of pollutants into plants is mainly through the leaves. There may also be a slight penetration through the stems and trunk. Before reaching the leaf, the pollutant will first have to pass through the "boundary layer" which corresponds to the layer of air not agitated in contact with it.

The thickness of this layer depends on the size and shape of the leaf, the presence of leaf hairs (or trichomes) and wind speed. Its thickness is in the order of a few tenths of a millimetre.

During the temporary presence of a pollutant in this boundary layer, many reactions are likely to occur because the incident pollutant will react with:

- An aqueous phase consisting of the water film present on the surface of the leaf as well as water bound to the polar groups of the cuticle;
- A lipid phase consisting of the waxes present within (intracuticular waxes) or located on the surface (epicuticular waxes) of the cuticle;
- A gaseous phase consisting of the components of the atmosphere and the emissions of the leaf.

Depending on the nature of the reactions that will or will not occur at the boundary layer, the concentration of the pollutant that will enter the plant can vary greatly. Some products of these reactions are even more phytotoxic than the pollutant itself.

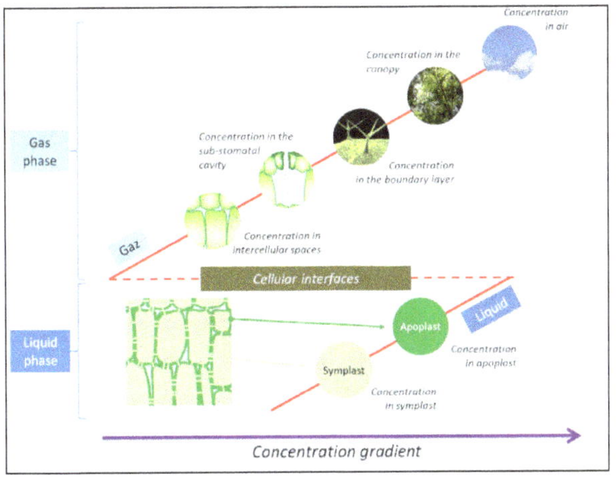

Schematic representation of the decreasing evolution of the concentration of a pollutant from the atmosphere to the inside of the leaf.

Gaseous pollutants enter the plant like other atmospheric gases (CO_2, Oxygen, etc.), mainly through stomata present on leaf surfaces. On the other hand, a large part of the organic pollutants will be absorbed mainly by the lipid structure of the cuticle. Only a small part will penetrate the leaf, then diffuse and react between and within the different internal compartments that constitute the apoplast and symplast.

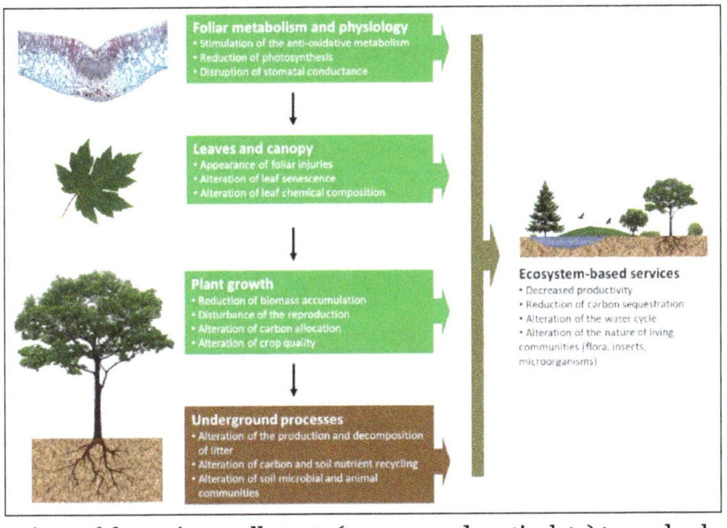

Behaviour of the various pollutants (gaseous and particulate) towards plants.

Particulate pollutants (organic or inorganic) are first captured by foliar surfaces (thanks to the micro-structure created by the presence of epicuticular waxes, trichomes, etc.), in a size range that is generally between 1 and 10 µm. In forests, this particulate deposition can vary between 280 and 1000 kg per hectare. Subsequently, meteorological conditions such as wind, sun and especially rain (leaching of leaves, dissolution of inorganic particles) influence the characteristics of this deposition. Thanks to the effectiveness of the cuticular barrier, organic or inorganic foliar deposition often causes only a slight penetration of pollutants into the leaves and therefore limit their physiological impacts.

After penetration, the physiological response of plants to air pollution will depend on the two actors involved: on the one hand, the characteristics of the plant and, on the other hand, the nature of the pollution.

Plant Reaction on Pollution

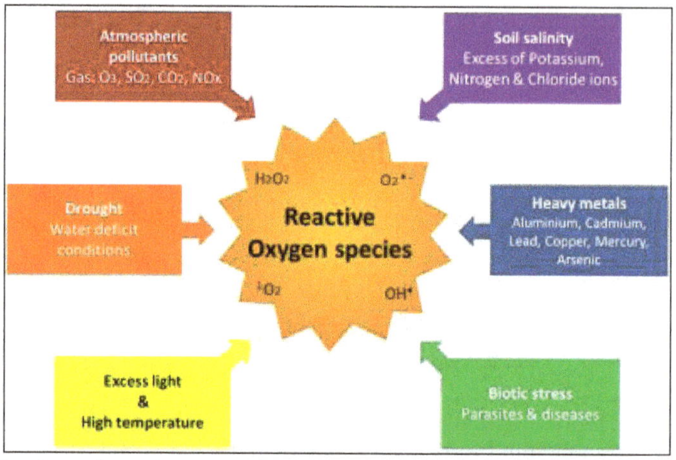

The production of reactive oxygen species (ROS) is a classic defence strategy when plants are subjected to a wide variety of stresses (drought, excess light, attack by pathogens, soil salinity, etc.). This is also the case when they are exposed to air pollutants.

Plants react to air pollution by producing reactive oxygen derivatives. After penetrating the leaves, and as for most biotic and abiotic stresses, the pollutants will first of all induce an oxidative stress with the production of free radicals (hydroxyl radicals) and reactive oxygen species (ROS) likely to cause damage at different levels. In particular, these ROS will have three main targets at the cell level: lipids (at the membrane level), proteins (at the amino acid level) and nucleic acids (adduct formation).

At the same time, the pollutant will cause specific stress related to its own physico-chemical characteristics:

- Thus, in the case of pollution by hydrofluoric acid (HF), there will be a disruption of the cellular metabolism of calcium (precipitation of calcium in the form of CaF_2).
- In the case of pollution with sulfur oxides (SO_2), the reducing properties of this gas will disturb the functioning of the photosynthetic apparatus (degradation of chlorophyll).
- Acid rain, on the other hand, causes mineral deficiencies leading to yellowing of the leaves due to rainwater leaching of the mineral elements Ca, K and Mg.

When facing these stresses, the traditional strategy of defending the plant aims to limit the absorption of the pollutant and increase its tolerance to it. It consists in implementing: (a) physical processes, i.e. closing stomata, falling leaves; and (b) chemical and biochemical processes.

These chemical and biochemical factors correspond to:

- Production of insoluble precipitates (formation of CaF_2 in the case of fluorine pollution);
- Detoxification by the reduced form emission of the pollutant (H_2S in the case of pollution by SO_2, NH_3 in the case of pollution by NOx);
- Enzymatic degradations by cytochromes P450 and a number of antioxidant enzymes. Non-enzymatic antioxidant compounds such as glutathione, vitamins E and C and carotenoids may also be involved.

When a "pollution stress" is installed, the plant will therefore set up (more or less quickly) processes that will be added to the pool of defence processes already present in the plant. Following the aggression, the plant's resistance to the pollutant will result from the combination of these various processes. For this reason, there is a specific scale of plant sensitivity for each pollutant and for each plant.

Visible and invisible damage – During low pollution and/or when the plant's defence systems are sufficient to limit the physiological impact of a pollutant, this resistance still has a physiological cost, which is characterized by decreases in size, in yield, etc. we then speak of "invisible damage".

During heavy pollution and/or when the plant's defence systems are not sufficient, irreversible damage appears such as cell death (leaf necrosis, among others). This is referred to as "visible damage" due to air pollution.

Plants react according to environmental conditions – The plant, like all biological systems, is sensitive at the same time to abiotic factors (temperature, humidity, light) and biotic factors (age, diseases, genotypes) in its environment. If diseases have a negative impact, other factors can have a positive impact on the plant's response to air pollution. Thus, drought leads to the closure of stomata, which protects the plant, while the increase in CO_2 promotes photosynthesis. The daily evolution of air pollution will also affect the response of plants. This is what field observations show:

- During hot weather, high temperatures lead to very high ozone concentrations in the air but at the same time to the closure of stomata. The result is a very low impact of this pollutant on vegetation during these periods. This was observed in the forests during the 2003 heat wave.
- During summer periods, ozone concentrations in the air around high altitude forests are high with slight day-night variations. As the high air humidity in these areas leads to a large opening of the stomata, a high impact of the ozone present is observed.
- On the other hand, at the level of lowland forests, air pollution is characterized by average ozone concentrations, this time with strong day-night variations. As the air humidity is lower in these areas, the opening of the stomata will be less important: for these two reasons, a lower impact of ozone is observed.

Pollutants

Target	Main pollutants	Effects on plants	Effets on humans via plants	Remarks
Field crops: cereals and oilseeds	Ozone	Yes Loss of yield (5 to 10%)	No	Losses masked by selection and fertilization
Vegetable, fruit and wine growing	Heavy metals Organic compounds	No	Yes Contamination of food chains	
Prairies	Ozone	Yes Loss of yield (5 to 10%)	No	Losses masked by increased CO_2 and nitrogen deposition
Production forests	Ozone	Yes Loss of yield (5 to 10%)	No	Losses masked by increased CO_2 and nitrogen deposition

Impact of air pollutants.

Depending on their chemical nature, pollutants are more or less phytotoxic. Laboratory experiments have classified the main air pollutants (at equal concentrations in air) in the following order of decreasing phytotoxicity:

Hydrofluoric acid (HF) > ozone (O_3) $_>$ sulfur dioxide (SO_2) > nitrogen dioxide (NO_2).

This classification is only given as an indication because there is a whole range of sensitivity of the different plants for each pollutant. For example, tobacco is very sensitive to ozone but not very affected by fluorine pollution.

In addition to the phytotoxicity of the pollutant, plants response will depend on the dose received (i.e. concentration x time). The dose is often calculated from the concentrations of pollutants in the atmosphere. It is the calculation of the flows of pollutants that have actually entered the leaf organs that provides the best information on pollution-damage relationships in vegetation.

Finally, at equal doses, the shorter the application time is, the greater is the pollutant impact. This "peak effect" is usually explained by the fact that, over short periods of time, the plant does not have time to start its defence systems.

Symptomatology

Symptomatology analyze the signs or manifestations (symptoms) expressed by plants in response to physiological disturbances induced by air pollutants.

This symptomatology is important because it can be used as a method of bio-monitoring of air quality with plants. This original method makes it possible to detect and estimate levels of air pollutants only by studying the visible (by observing necrosis) or invisible (by biochemical analyses) disturbances they cause on plants.

As the pollutants present around plants are very numerous, it is interesting to classify them, for symptomatological observations, according to the extent of their impact zone: local, regional or global.

Local Pollutants

Local pollutants will have, at the plant level, impacts on at most a few tens of kilometres around their emission sources. These are mainly nitrogen compounds emitted directly from pollution sources (primary pollutants), mainly NOx (from transport) and NH_3 (from agriculture and transport). These nitrogen compounds are paradoxical pollutants as they are not very phytotoxic but have a strong impact on vegetation:

- They have a beneficial fertilizing effect by promoting growth in the first place;
- But they have a negative impact over time by causing eutrophication of ecosystems, mineral deficiencies, effects on biodiversity (nitrophilic plants are favoured) and a decrease in resistance to various stresses.

Other local pollutants in the air are particulate deposits. It should be remembered that the vast majority of air pollution sources are both emitters of gas and dust (particles). Particulate deposits are composed of an inorganic fraction (heavy metals), but also an organic fraction (PAH, etc). They include, among other things:

- Transport related pollutants. These are organic compounds such as BTX (VOCs in exhaust gases), or inorganic compounds such as platinoids: Pt, Rh, Pd (catalytic converters), titanium (aircraft engines), etc.
- Pollutants related to agriculture following the transfer of plant protection products such as pesticides (herbicides, fungicides, insecticides) into the air during spraying.
- Emissions from incineration facilities with in particular organic compounds such as dioxins, furans, PCBs, etc.

All these different particulate pollutants have little or no impact on vegetation, but they cause contamination of the food chains of humans and animals via plants.

For some local pollutants, the impact may be more pronounced. The massive use of more or less biodegradable detergents discharged into the sea leads to water and then air pollution (from the formation by winds of spray loaded with detergents present on the surface). The deposition of these surfactants on the leaves will then promote the penetration of salt into the plants, causing their subsequent decline and death. Impacts of this particular pollution can be observed on the edges of certain coastal forests around the Mediterranean.

Regional Pollutants

Regional pollutants can have impacts over several hundred kilometres around their emission sources. They mainly include acid deposition, with mainly the presence of H_2SO_4 and HNO_3 in wet or solid deposition. They are secondary pollutants because they are the result of interactions between primary pollutants (SO_2, nitrogen compounds) and ozone.

Impact of ozone on tobacco leaves BEL W3.

Acid deposition has a low impact on plants: yellowing of needles and decrease in tree vitality due to leaching of the ions they induce. On the other hand, ozone is a very phytotoxic gas due to its direct and highly oxidizing effects on the plant's various physiological processes (photosynthesis, respiration, etc.). It is the most worrying pollutant currently affecting vegetation and ecosystems because it causes yield losses of up to 5 to 10%, and the appearance of leaf necrosis now visible in natural environments. But it also has indirect effects on vegetation because it is a greenhouse gas linked to climate change.

It is now considered that 90% of the yield losses related to air pollution in the plant world come from ozone. However, this observation must be put into perspective for the growth of vegetation in anthropized ecosystems (field crops, etc.): the negative effects of ozone are very often masked by the positive effects on photosynthesis of the increase in CO_2 in the atmosphere (280 ppm before the industrial era compared to more than 400 ppm currently).

Following this significant impact of ozone on vegetation and in order to detect and assess its effects, numerous lists have been drawn up indicating the sensitivity of plants present in different natural or anthropogenic environments (forests, grasslands, field crops, etc.) to ozone, depending on the climatic zones encountered (Western and Central Europe, Mediterranean coast of Europe, etc) .

Global Pollutants

Global pollutants have global impacts. They mainly include CO_2, which is a pollutant linked to the massive use of fossil fuels by transport and industry. CO_2 is a paradoxical pollutant, which will have direct beneficial effects on plant growth via its essential role in photosynthesis. But at the same time, it has indirect harmful effects on plants through the greenhouse effect and the resulting climate disturbances. Other global pollutants include:

- Methane or CH_4, a gas produced by digesting herbivorous animals, anaerobic environments such as rice fields;
- CFCs and HFCs (used in refrigerators or as solvents);
- N_2O (from the massive use of fertilizers in agriculture);
- Methyl bromide (used as a disinfectant in sericulture soils).

All these other gases only have an indirect effect on vegetation via their roles in the greenhouse effect. They are also, with the exception of CH_4, ozone-depleting gases with the possibility of another negative impact on plants as a result of increased solar UV-B fluxes reaching the ground.

References

- 3-how-do-soil-organisms-affect-plants, soils-are-alive: soilhealth.com, Retrieved 17 May, 2019
- Agri-min-nutri-def-symptoms, agriculture: tnau.ac.in, Retrieved 19 April, 2019
- Temperature-extremes-effect-on-plant-growth-and-development: researchgate.net, Retrieved 5 February, 2019
- Toxic-chemicals, plant-disease, science: britannica.com, Retrieved 26 July, 2019
- Impact-air-pollutants-on-vegetation, life: encyclopedie-environnement.org, Retrieved 21 May, 2019

4

Common Plant Diseases

The different plant diseases are broadly classified into fungal, bacterial and viral plant diseases. Some of the fungal diseases are fungal leaf spots, rice blast fungus, botrytis cinerea, stem rust, etc. Aster yellows, bacterial wilt, bacterial spot, bacterial blight, etc. are some of the bacterial diseases. The most common viral diseases are phyllody and barley yellow dwarf. This chapter discusses all these plant diseases in detail.

FUNGAL DISEASES

Fungal Leaf Spots

Pathogen-caused leaf spot diseases, particularly those of stone fruit trees and such vegetables as tomato, pepper and lettuce are of two types, those caused by bacteria and those caused by fungus. Leaf spotting of either kind is generally similar in appearance and effect. Prevention and treatment of both kinds often involve the same practices.

Symptoms of Fungal Leaf Spots

Infected plants have brown or black water-soaked spots on the foliage, sometimes with a yellow halo, usually uniform in size. The spots enlarge and will run together under wet conditions. Under dry conditions the spots have a speckled appearance. As spots become more numerous, entire leaves may yellow, wither and drop. Members of the Prunus family (stone fruits, including cherry, plum, almond, apricot and peach) are particularly susceptible to bacterial leaf spot. The fruit may appear spotted or have sunken brown areas. Bacterial leaf spot will also attack tomato and pepper crops in vegetable gardens.

Fungal leaf spot attacks lettuce and can also occur on brassicas and other vegetables including such as cabbage, cauliflower, Chinese cabbage, broccoli, Brussels sprouts, kohlrabi, kale, turnip and rutabaga.

Bacterial leaf spot will also infect some annual and perennial flowering plants including geraniums, zinnias, purple cone flowers and black-eyed Susan. Fungal leaf spot will infect aspen and poplar trees. Leaf spot will also cause problems for strawberry plants.

Both types of leaf spot are most active when there is plenty of moisture and warm temperatures. During the summer months, especially if plants are watered by overhead sprinklers, sufficient moisture may be present for infection when the bacteria are splashed or blown on to leaves. Wind and rain transmit the bacteria to plants.

This disease overwinters in the soil around infected plants as well as on garden debris and seeds. It will also remain in the twig cankers, leaves, stems and fruit of infected trees.

Control

- When selecting fruit trees, choose resistant varieties if possible.
- Keep the soil under the tree clean and rake up fallen fruit.
- Use a thick layer of mulch to cover the soil after you have raked and cleaned it well. Mulch will reduce weeds and prevent the disease pathogen from splashing back up onto the leaves.
- Prune or stake plants to improve air circulation. Make sure to disinfect your pruning equipment after each cut.
- Leaf spot among vegetables is most often introduced through infected seed or transplants.

Magnaporthe Oryzae/Rice Blast Fungus

Magnaporthe oryzae or rice blast fungus, causes rice blast, the most important fungal rice disease in the world. Taxonomic research indicated that M. oryzae is distinct from M. grisea, a species that is morphologically indistinguishable from M. oryzae, a species that affects crabgrass (Digitaria). M. oryzae, infects millet and other grasses but primarily rice, the most economically important host. Leaf blast may result in significant economic loss; however, it is the panicle blast symptom that has the greater economic impact.

The biology of M. oryzae has been studied as a model to further understand fungal/plant interactions. Rice blast fungus is a hemiobiotrophic pathogen meaning that initially, the fungus establishes a biotrophic relationship with its host (i.e. invades a few cells, steals nutrients, but does not kill host cells). Eventually, the fungus becomes necrotrophic, destroying plant tissue. Infection starts when three-celled conidia from the pathogen adhere to the host's surface. Initial attachment of conidia and other fungal structures is mediated by mucilage. Upon germination of the conidia, an appressorium is produced. This develops a penetration peg that breaks through the cell wall giving rise to proliferation of hyphae. Cell to cell movement is achieved through plasmodesmata. After colonization of the host and infection of aerial parts, lesions are formed and sporulation takes place. The fungus produces new conidia that disperse through the air and serve as secondary inoculum, giving continuity to the epidemic. Fungal resting structures, hyphae inside crop residues and conidia, allow the survival of the pathogen off season. These are the source of primary inoculum that enables the disease to reinitiate.

Symptoms and Signs of Magnaporthe Oryzae/Rice Blast Fungus

Symptoms can occur on different parts of the plant.

- Leaf blast: Symptoms on the leaves vary with the level of resistance of the cultivar. On a susceptible cultivar, water soaked areas with a diamond shape develop from areas that start as gray to green in color. A dark green color develops on the border of the lesions. As they expand and mature, they turn tan to brown and the border becomes necrotic. The diamond-shape lesions on the leaves, are a key characteristic for the quick identification of the disease in the field.

- Collar rot: Rice plant collars appears necrotic at the point where the stem sheath and the leaf come together. The infection by on the plant's neck causes it to rot and can result in a lack of seed filling or collapse of the entire panicle (i.e. panicle blast). As the fungus infects the seed-forming panicle, the seeds may become infected. If the pedicels are infected, seeds may not be produced at all. Brown spots or diamond shape lesions can develop on infected seed.

Ecology and Spread

Infected grasses, volunteers, and crop residues are usually the most important sources of inoculum. Infected seed can also be a source of inoculum. After infection and colonization by the pathogen, conidia are produced on the lesions. Each lesion can produce thousands of spores which are readily spread by air. The high inoculum potential combined with several cycles of the pathogen within one season, result in the infection of several stages of the crop. The pathogen is favored by warm and humid conditions with an optimum temperature of 24 °C and >12 hours of high moisture. Release of the conidia is favored by darkness and wind, which can spread the inoculum up to 230 meters from the source of origin under tropical conditions. Symptoms may develop as quickly as four days after infection on young plants.

Geographic Distribution

The pathogen is cosmopolitan and found in the tropics and all temperate areas where rice is cultivated. Rice cropping systems (i.e. upland, lowland, irrigated and rainfed) can all be severely affected by rice blast.

Management

Management of the disease should not rely solely on chemical control or host resistance. It must be achieved through the combination of multiple management strategies.

- Adopt a forecasting system such as BLASTAM to monitor possible outbreaks of leaf blast.
- Start with certified seed or pre-treat your seed with hot water or fungicides.
- Use a mixture of resistant varieties and include susceptible ones in your crop to avoid breaking of host resistance.
- Practice proper sanitation measures to a reduce inoculum present in the field.
- Carry out crop rotation, intercropping, and/or fallow.
- Actively manage weeds and destroy volunteers to reduce number of hosts and survival of the pathogen.
- Modify plant density so to avoid close spacing since this favors warm and humid conditions ideal for blast. Varieties with high tillering should be planted more separately compared to low tillering varieties.
- Provide a balanced nutrition to the rice crop.
- Limit fungicide applications to loss-susceptible stages.

- If possible, develop a wide area management strategy for the region to disrupt the continuous pathosystem created by staggered rice plantations within a region.
- Consult your local extension specialist for legal and efficacious fungicide products available in your state.Remember, the label is the law and the product applicator is responsible for reading and following all chemical labeling.

Diagnostic Procedures

- Field identification: Symptoms of rice blast include diamond-shaped white to gray-green lesions on the leaves. These lesions are usually 1.0-1.5 cm in length with a width of 0.3-0.5 cm. Partly unfilled panicles are also a symptom of the disease. The panicles can exhibit blast, collar, and neck rot. For an accurate diagnosis of the disease, leaf, neck nodes, and leaf collars presenting a variety of symptoms (i.e. young and mature lesions) should be sampled.
- Lab identification: Symptomatic samples can be placed into a moist chamber to induce sporulation. Samples should be incubated for 48 hours at room temperature before observation with a light microscope. Additionally, samples can be plated on different media following surface sterilization of the symptomatic tissue. Media used for isolation of the pathogen includes Potato Destroxe Agar (PDA – general medium), cornmeal, rice straw, and oatmeal agar. These media promote sporulation of Pyricularia oryzae. Compare the conidia obtained from moist chamber and media plates. The conidia are usually light colored (i.e. hyaline) or lightly pigmented with an olive-green color. These have a pear-like shape (i.e. pyriform), present two septa (mostly), and have a basal appendage. The conidiophores can be observed emerging directly from the plant tissue bearing up to 20 conidia.

Botrytis Cinerea

Botrytis cinerea is a necrotrophic fungus that affects many plant species, although its most notable hosts may be wine grapes. In viticulture, it is commonly known as "botrytis bunch rot"; in horticulture, it is usually called "grey mould" or "gray mold".

The fungus gives rise to two different kinds of infections on grapes. The first, grey rot, is the result of consistently wet or humid conditions, and typically results in the loss of the affected bunches. The second, noble rot, occurs when drier conditions follow wetter, and can result in distinctive sweet dessert wines, such as Sauternes. The fungus is usually referred to by its anamorph (asexual form) name, because the sexual phase is rarely observed. The teleomorph (sexual form) is an ascomycete, *Botryotinia fuckeliana*, also known as *Botryotinia cinerea.*

Hosts

The disease, gray mold, affects more than 200 dicotyledonous plant species and a few monocotyledonous plants found in temperate and subtropical regions. Serious economic losses can be a result of this disease to both field and greenhouse grown crops. The causal agent, *Botrytis cinerea* can infect mature or senescent tissues, plants prior to harvest, or seedlings. There is a wide variety of hosts infected by this pathogen including protein crops, fiber crops, oil crops, and horticultural crops. Horticultural crops include vegetables (examples are chickpeas, lettuce, broccoli, and beans) and small fruit crops (examples are grape, strawberry, and raspberry), these are most severely affected and devastated by gray mold. Plant organs affected include fruits, flowers, leaves, storage organs, and shoots.

Biology

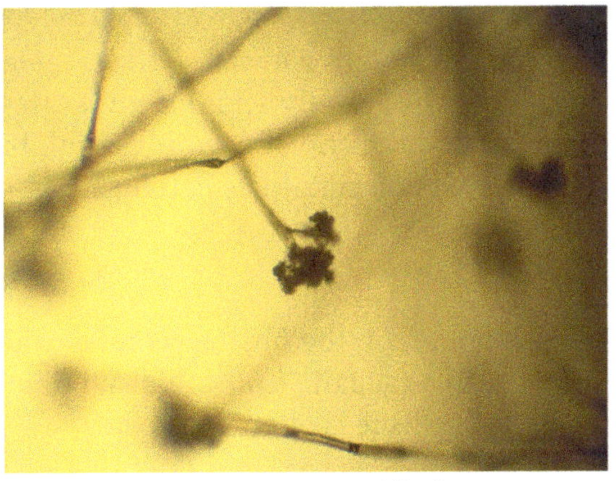

A *Botrytis cinerea* conidiophore.

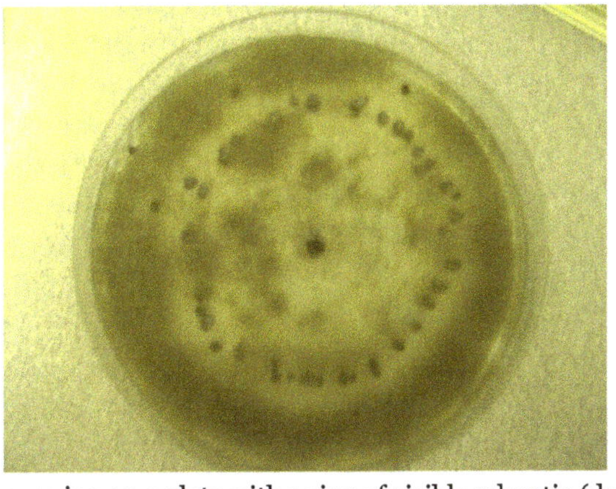

Botrytis cinerea growing on a plate with a ring of visible sclerotia (dark brown balls).

Botrytis cinerea is characterized by abundant hyaline conida (asexual spores) borne on grey, branching tree-like conidiophores. The fungus also produces highly resistant sclerotia as survival structures in older cultures. It overwinters as sclerotia or intact mycelia, both of which germinate in spring to produce conidiophores. The conidia, dispersed by wind and by rain-water, cause new infections. Different *Botrytis cinerea* strains show considerable genetic variability (polyploidy). *Gliocladium roseum* is a fungal parasite of *Botrytis cinerea.*

Environment

Gray mold favors moist, humid, and warm environmental conditions between 18.3-23-9 °C (65-75 °F). Temperature, relative humidity, and wetness duration produce a conducive environment that is favorable for inoculation of mycelium or conidia. Controlled environments, such as crop production greenhouses, provide the moisture and high temperatures that favor the spreading and development of the pathogen *Botrytis cinerea.*

Standing water on plant leaf surfaces provides a place for spores to germinate. Humid conditions can result from improper irrigation practice, plants placed too close together, or the structure of the greenhouse not allowing for efficient ventilation and air flow. Ventilation at night significantly reduces the incidence of gray mold.

Melanized sclerotium allows *Botrytis cinerea* to survive for years in the soil. Sclerotia and the asexual conidia spores contribute to the widespread infection of the pathogen.

A low pH is preferred by the gray mold to perform well. *Botrytis cinerea* can acidify its environment by secreting organic acids, like oxalic acid. By acidifying its surroundings, cell wall degrading enzymes (CWDEs) are enhanced, plant-protection enzymes are inhibited, stomatal closure is deregulated, and pH signaling is mediated to facilitate its pathogenesis.

Viticulture

In the *Botrytis* infection known as noble rot, the fungus removes water from the grapes, leaving behind a higher percent of solids, such as sugars, fruit acids and minerals. This results in a more intense, concentrated final product. The wine is often said to have an aroma of honeysuckle and a bitter finish on the palate.

A distinct fermentation process initially caused by nature, the combination of geology, climate and specific weather led to the particular balance of beneficial fungus while leaving enough of the grape intact for harvesting.

Botrytis complicates winemaking by making fermentation more complex. *Botrytis* produces an anti-fungal that kills yeast and often results in fermentation stopping before the wine has accumulated sufficient levels of alcohol. Makers of fine German dessert wines have been known to take fermenting tubs of wine into their homes to nurture the yeast through the night to assure that the alcohol level reaches legal minimums for the product to be called wine.

Botrytis cinerea on Riesling grapes.

Botrytis bunch rot is another condition of grapes caused by *Botrytis cinerea* that causes great losses for the wine industry. It is always present on the fruitset, however, it requires a wound to start a bunch rot infection. Wounds can come from insects, wind, accidental damage, etc. To control botrytis bunch rot there are a number of fungicides available on the market. Generally, these should be applied at bloom, bunch closure and veraison (the most important being the bloom application). Some winemakers are known to use the German method of fermentation and prefer having a 5% bunch rot rate in their grapes and will usually hold the grapes on the vine a week longer than normal.

Horticulture

Botrytis cinerea affects many other plants. It is economically important on soft fruits such as strawberries and bulb crops. Unlike wine grapes, the affected strawberries are not edible and are discarded. To minimize infection in strawberry fields, good ventilation around the berries is important to prevent moisture being trapped among leaves and berries. A number of bacteria have been proven to act as natural antagonists to *B. cinerea* in controlled studies.

In greenhouse horticulture, *Botrytis cinerea* is well known as a cause of considerable damage in tomatoes.

The infection also affects rhubarb, snowdrops, white meadowfoam, western hemlock, Douglas-fir and cannabis. Potassium bicarbonate-based fungicide has been proven to cure and prevent powdery mildew, blackspot, downy mildew, blights, molds and other plant diseases, such as Botrytis cinerea.

Mycoviruses of Botrytis Cinerea

Botrytis cinerea not only infects plants, it also hosts several mycoviruses itself.

A range of phenotypic alterations due to the mycoviral infection have been observed from symptomless to mild impact, or more severe phenotypic changes including reduction in pathogenicity, growth/suppression of mycelia, sporulation and sclerotia production, formation of abnormal colony sectors and virulence.

Management

Botrytis cinerea can be managed through cultural, chemical, and biological practices.

There are no resistant species to the gray mold rot. Gray mold can be culturally controlled by monitoring the amount and timing of fertilizer applications to reduce the amount of fruit rot. Excessive application of nitrogen will increase the incidence of disease while not improving yields.

Not planting cultivars that have an upright or dense growth habit can reduce disease as these limit airflow and are favorable for the pathogen. Spacing of plants so they are not touching will increase airflow allowing the area to dry out and reduce the spread of disease. Pruning or purposeful removal of diseased, dead, or overgrown limbs on a regular schedule can also help to improve air movement.

Sanitation by removing dead or dying plant tissue in the fall will decrease inoculum levels as there is no debris for the sclerotium or mycelia to overwinter. Removing debris in the spring will remove inoculum from the site. Disposal of berries during harvest that have signs and symptoms of gray mold will reduce inoculum for the following year.

Biochar, a form of charcoal, can be applied as a soil amendment to strawberry plants to reduce the severity of the fungal disease by stimulating defense pathways within the plant.

Gray mold can be chemically controlled with well-timed fungicide applications starting during the first bloom. Timing can reduce the chance of resistance and will save on costs.

Biological controls or microbial antagonists used for disease suppression, have been successfully used in Europe and Brazil in the form of fungi-like *Trichoderma harzianium* Rifai and *Gliocladium roseum* Bainier. *Trichoderma* species especially, have been shown to control gray mold.

Stem Rust

The stem, black, and cereal rusts are caused by the fungus *Puccinia graminis* and are a significant disease affecting cereal crops. Crop species that are affected by the disease include bread wheat, durum wheat, barley and triticale. These diseases have affected cereal farming throughout history. The life cycle of Puccinia (also called the "rust cycle") was discovered by Prof. K.C. Mehta. Since the 1950s, wheat strains bred to be resistant to stem rust have become available. Fungicides effective against stem rust are available as well.

In 1999 a new virulent race of stem rust was identified that most current wheat strains show no resistance against. The race was named TTKSK (e.g. isolate Ug99), named after the country where it was identified (Uganda) and the year of its discovery (1999). It spread to Kenya, then Ethiopia, Sudan and Yemen, and is becoming more virulent as it spreads. An epidemic of stem rust on wheat caused by race TTKSK is currently spreading across Africa, Asia and the Middle East and is causing major concern due to the large numbers of people dependent on wheat for sustenance. Scientists are working on breeding strains of wheat that are resistant to UG99. However, wheat is grown in a broad range of environments. This means that breeding programs would have extensive work remaining to get resistance into regionally adapted germplasms even after resistance is identified.

An outbreak of another virulent race of stem rust, TTTTF, took place in Sicily in 2016, suggesting that the disease is returning to Europe. Comprehensive genomic analysis of *Puccinia graminis* combined with plant pathology and climate data has pointed out the potential of the re-emergence of stem wheat rust in UK.

Biology

There is considerable genetic diversity within the species *P. graminis*, and several special forms, *forma specialis*, which vary in host range have been identified.

- *Puccinia graminis* f. sp. *avenae*, oat
- Puccinia graminis f. sp. dactylis "Puccinia graminis" f. sp. "dactylis"
- *Puccinia graminis* f. sp. *hordei*, barley

- *Puccinia graminis* f. sp. *lolii*
- *Puccinia graminis* f. sp. *poae*
- *Puccinia graminis* f. sp. *secalis*, rye, barley
- *Puccinia graminis* f. sp. *tritici*, wheat, barley

P. graminis is a member of the phylum Basidiomycota within the kingdom Fungi. The characteristic rust color on stems and leaves is typical of a general stem rust as well as any variation of this type of fungus. Different from most fungi, the rust variations have five spore stages and alternate between two hosts. Wheat is the primary host, and barberry is the alternate host.

There are multiple pathotypes (including QCC and MCC) affecting barley, within *forma specialis tritici*.

Model of a spore of *puccinia graminis*, late 19th century.

Pathology

The stem rust fungus attacks the parts of the plant that are above ground. Spores that land on green wheat plants form a pustule that invades the outer layers of the stalk. Infected plants produce fewer tillers and set fewer seed, and in cases of severe infection the plant may die. Infection can reduce what is an apparently healthy crop about three weeks before harvest into a black tangle of broken stems and shriveled grains by harvest.

Stem rust of cereals causes yield losses in several ways:

- Fungus absorbs nutrients that would otherwise be used for grain development
- Pustules break through epidermis, which disrupt the plant's control of transpiration and can lead to desiccation and infection by other fungi
- Interference with plant vascular tissue leads to shriveled grains
- The fungus weakens the stems, which can lead to lodging (falling over). In severe cases lodging can make mechanical harvesting impossible

Signs and Symptoms

On Wheat

Stem rust on wheat is characterized by the presence of uredinia on the plant, which are brick-red, elongated, blister-like pustules that are easily shaken off. They most frequently occur on the leaf sheaths, but are also found on stems, leaves, glumes and awns. On leaves they develop mostly on the underside but may penetrate to the upperside. On leaf sheaths and glumes pustules rupture the epidermis, giving a ragged appearance.

Wheat infected leaves with stem rust pathogen with a specific resistance gene.

Towards the end of the growing season black telia are produced. For this reason stem rust is also known as 'black rust'. The telia are firmly attached to the plant tissue.

The site of infection is a visible symptom of the disease.

On Barberry

Pycnia appear on barberry plants in the spring, usually in the upper leaf surfaces. They are often in small clusters and exude pycniospores in a sticky honeydew. Five to ten days later, cup-shaped structures filled with orange-yellow, powdery aeciospores break through the lower leaf surface. The aecial cups are yellow and sometimes elongate to extend up to 5 mm from the leaf surface.

Life Cycle

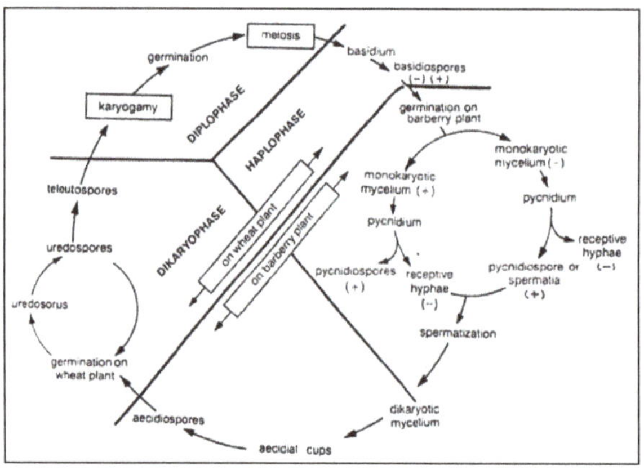

Life cycle of *Puccinia graminis.*

Like other *Puccinia* species, *P. graminis* is an obligate biotroph (it colonizes living plant cells) and has a complex life cycle featuring alternation of generations. The fungus is heteroecious, requiring two hosts to complete its life cycle – the cereal host and the alternate host. There are many species in *Berberis* and *Mahonia* that are susceptible to stem rust, but the common barberry is considered to be the most important alternate host. P. graminis *is macrocyclic* (exhibits all five of the spore types that are known for rust fungi).

P. graminis can complete its life cycle either with or without barberry (the alternate host).

Life Cycle on Barberry

Due to its cyclical nature, there is no true 'start point' for this process. Here, the production of urediniospores is arbitrarily chosen as a start point.

Urediniospores are formed in structures called uredinia, which are produced by fungal mycelia on the cereal host 1–2 weeks after infection. The urediniospores are dikaryotic (contain two un-fused, haploid nuclei in one cell) and are formed on individual stalks within the uredinium. They are spiny and brick-red. Urediniospores are the only type of spores in the rust fungus life cycle that are capable of infecting the host on which they are produced, and this is therefore referred to as the 'repeating stage' of the life cycle. It is the spread of urediniospores that allows infection to spread from one cereal plant to another. This phase can rapidly spread the infection over a wide area.

Towards the end of the cereal host's growing season, the mycelia produce structures called telia. Telia produce a type of spore called teliospores. These black, thick-walled spores are dikaryotic. They are the only form in which *Puccinia graminis* is able to overwinter independently of a host.

Each teliospore undergoes karyogamy (fusion of nuclei) and meiosis to form four haploid spores called basidiospores. This is an important source of genetic recombination in the life cycle. Basidiospores are thin-walled and colourless. They cannot infect the cereal host, but can infect the alternative host (usually barberry). They are usually carried to the alternative host by wind.

Once basidiospores arrive on a leaf of the alternative host, they germinate to produce a haploid mycelium that directly penetrates the epidermis and colonises the leaf. Once inside the leaf the mycelium produces specialised infection structures called pycnia. The pycnia produce two types of haploid gametes, the pycniospores and the receptive hyphae. The pycniospores are produced in a sticky honeydew that attracts insects. The insects carry pycniospores from one leaf to another. Splashing raindrops can also spread pycniospores. A pycniospore can fertilise a receptive hypha of the opposite mating type, leading to the production of a dikaryotic mycelium. This is the sexual stage of the life cycle and cross-fertilisation provides an important source of genetic recombination.

This dikaryotic mycelium then forms structures called aecia, which produce a type of dikaryotic spores called aeciospores. These have a worty appearance and are formed in chains – unlike the urediniospores that are spiny and are produced on individual stalks. The chains of aeciospores are surrounded by a bell-like enclosure of fungal cells. The aeciospores are able to germinate on the cereal host but not on the alternative host (they are produced on the alternative host, which is usually barberry). They are carried by wind to the cereal host where they germinate and the germ tubes penetrate into the plant. The fungus grows inside the plant as a dikaryotic mycelium. Within 1–2 weeks the mycelium produces uredinia and the cycle is complete.

Life Cycle without Barberry

Since the urediniospores are produced on the cereal host and can infect the cereal host, it is possible for the infection to pass from one year's crop to the next without infecting the alternate host (barberry). For example, infected volunteer wheat plants can serve as a bridge from one growing season to another. In other cases the fungus passes between winter wheat and spring wheat, meaning that it has a cereal host all year round. Since the urediniospores are wind dispersed, this can occur over large distances. Note that this cycle consists simply of vegetative propagation urediniospores infect one wheat plant, leading to the production of more urediniospores that then infect other wheat plants.

Spore Dispersal

Puccinia graminis produces all five of the spore types that are known for rust fungi.

Spores are typically deposited close to the source, but long-distance dispersal is also well documented. The following three categories of long-distance dispersal are known to occur:

- Extremely long-distance dispersal: This can occur unassisted (the robust nature of the spores allows them to be carried long distances in the air and then deposited by rain-scrubbing) or assisted (typically on human clothing or infected plant material that is transported between regions). This type of dispersal is rare and is very difficult to predict.
- Step-wise range expansion: This is probably the most common mode of long-distance dispersal and usually occurs within a country or region.
- Extinction and recolonisation: This occurs in areas that have unsuitable conditions for year-round survival of *Puccinia graminis* – typically temperate regions where hosts are absent during either the winter or summer. Spores overwinter or oversummer in another region and then recolonise when conditions are favorable.

Wheat Stem Rust Resistance Genes

A number of stem rust resistance genes (Sr genes) have been identified in wheat. Some of them arose in bread wheat (e.g. *Sr5* and *Sr6*), while others have been bred in from other wheat species (e.g. *Sr21* from *T. monococcum*) or from other members of the tribe *Triticeae* (e.g. *Sr31* from rye and *Sr44* from *Thinopyrum intermedium*).

None of the Sr genes provide resistance to all races of stem rust. For instance many of them are ineffective against the Ug99 lineage. Notably Ug99 has virulence against *Sr31*, which was effective against all previous stem rust races. Recently, a new stem rust resistance gene *Sr59* from *Secale cereale* was introgressed into wheat, which provides an additional asset for wheat improvement to mitigate yield losses caused by stem rust. Singh et al. provide a list of known Sr genes and their effectiveness against Ug99.

Fusarium Ear Blight

Fusarium ear blight (FEB) (also called Fusarium head blight, FHB, or scab), is a fungal disease of cereals, including wheat, barley, oats, rye and triticale. FEB is caused by a range of *Fusarium* fungi, which infects the heads of the crop, reducing grain yield. The disease is often associated withcontamination by mycotoxins produced by the fungi already when the crop is growing in the field. The disease can cause severe economic losses as contaminated grain cannot be sold for food or feed.

Symptom on wheat caused by *F. graminearum* (right:inoculated, left:non-inoculated).

Causal Organism

Fusarium ear blight is caused by several species of *Fusarium* fungi, belonging to the Ascomycota. The most common species causing FEB are:

- *Fusarium avenaceum* (teleomorph: *Gibberella avenacea*)
- *Fusarium culmorum*
- *Fusarium graminearum* (teleomorph: *Gibberella zeae*)
- *Fusarium poae*
- *Microdochium nivale* (teleomorph: *Monographella nivalis*, formerly *Fusarium nivale*)

Fusarium graminearum is considered the most important causal organism.

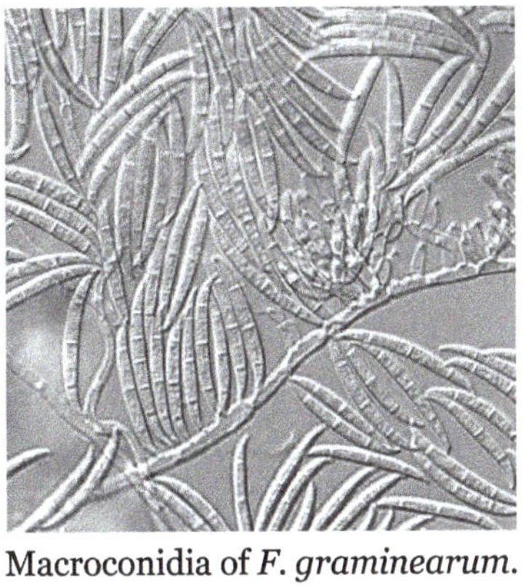

Macroconidia of *F. graminearum*.

Fusarium species causing FEB can produce several types of spores. The asexual stage of the fungus produces spores called macroconidia. Some *Fusarium* fungi have a more complex life cycle including a sexual stage, for example *F. graminearum*. In the sexual stage the fungus produces spores called ascospores. The sexual stage form fruiting bodies called perithecia, in which ascospores are formed in a sac known as an ascus (plural asci). Some species including *F. culmorum* produes resistant chlamydospores which can survive for a long time in the soil.

Disease Cycle and Epidemiology

Fusarium fungi can overwinter as saprotrophs in the soil or on crop debris that can serve as inoculum for the following crop. The fungus can also spread via infected seed. The presence of *Fusarium* fungi on crop debris or seed can cause *Fusarium* seedling blight and foot and root rot. Later, infection of the heads can occur with spores spreading by rain splash from infected crop residues. Another major infection route is airborne inoculum as spores can travel long distances with the wind. The cereal crop is most susceptible at flowering and the probability of infection rises with high moisture and humidity at flowering.

Symptoms

In wheat, *Fusarium* infects the head (hence the name "Fusarium head blight") and causes the kernels to shrivel up and become chalky white. Additionally, the fungus can produce mycotoxins that further reduce the quality of the kernel.

Infected florets (especially the outer glumes) become slightly darkened and oily in appearance. Conidiospores are produced in sporodochia, which gives the spike a bright pinkish color. Infected kernels may be permeated with mycelia and the surface of the florets totally covered by white, matted mycelia.

Mycotoxins

Fusarium species associated with FEB produce a range of mycotoxins—fungal secondary metabolites with toxic effects on animals. One mycotoxin can be produced by several *Fusarium* species, and one species can produce several mycotoxins. Important *Fusarium* mycotoxins include:

- Deoxynivalenol (DON) produced by *F. graminearum* and *F. culmorum.*
- Zearalenone (ZEN) produced by *F. graminearum* and *F. culmorum.*
- HT-2 and T-2 produced by *F. langsethiae.*

Fusarium toxins have negative effects on the immune, gastrointestinal and reproductive systems of animals. DON is a protein synthesis inhibitor, also called vomitoxin, due to its negative effects on feed intake in pigs. Pigs are the most sensitive to DON, while ruminant animals such as cattle have higher tolerance.

Many countries monitor *Fusarium* mycotoxins in grain to limit negative health effects.

Control Measures

Resistant Cultivars

Resistant cultivars could be the most efficient method to control Fusarium ear blight. Resistance breeding involves screening of plant lines subjected to artificial inoculation with *Fusarium.* Plant lines having reduced fungal growth and low levels of seed mycotoxin contamination are selected for additional breeding trials. In parallel, genetic markers associated with resistance are screened for, so called marker-assisted selection. Fusarium ear blight resistance is a complex trait, involving several genes, and is dependent of interaction with the environment.

Fusarium ear blight resistance has been identified in wheat cultivars from Asia. However, the challenge is to combine resistant material with other desirable traits such as high yield and adaptation to different growing areas.

Agricultural Practices

Several agricultural practices affect the risk of FEB. One of the major infection routes are infected crop residues from the previous crop where both the quality and quantity are important. Crop residues from susceptible crops such as cereals increase the risk of FEB in the following crop. Maize has been associated with especially high risk. Reduced soil tillage can also increase the risk of FEB. The amount of crop residues can be reduced by ploughing, where residues are incorporated in the soil where they decompose faster. High nitrogen application has also been associated with increased risk of *Fusarium* infection. Preventive agricultural practices may be less effective if a lot of airborne inoculum is present in the area.

Chemical Control

Fungicides can provide partial control of FEB but the effects may be variable. The type and timing of fungicide application is important as non-optimal applications may even increase *Fusarium* infection.

Biological Control and Integrated Management

Research has also been put into development on biological control strategies based on bacteria and fungi for example, *Bacillus* and *Cryptococcus* species.

For FEB no control measure is complemetely effective and integrated management involving several control strategies such as preventive measures, disease monitoring and chemical control is necessary. Disease forecasting models have been developed to assess the risk of FEB depending on weather conditions.

Powdery Mildew

Powdery mildew is a fungal disease that affects a wide range of plants. Powdery mildew diseases are caused by many different species of fungi in the order Erysiphales, with *Podosphaera xanthii* (a.k.a. *Sphaerotheca fuliginea*) being the most commonly reported cause. *Erysiphe cichoracearum* was formerly reported to be the primary causal organism throughout most of the world. Powdery mildew is one of the easier plant diseases to identify, as its symptoms are quite distinctive. Infected plants display white powdery spots on the leaves and stems. The lower leaves are the most affected, but the mildew can appear on any above-ground part of the plant. As the disease progresses, the spots get larger and denser as large numbers of asexual spores are formed, and the mildew may spread up and down the length of the plant.

Powdery mildew grows well in environments with low humidity and moderate temperatures. Greenhouses provide an ideal moist, temperate environment for the spread of the disease. This causes harm to agricultural and horticultural practices where powdery mildew may thrive in a greenhouse setting. In an agricultural or horticultural setting, the pathogen can be controlled using chemical methods, bio organic methods, and genetic resistance. It is important to be aware of powdery mildew and its management as the resulting disease can significantly reduce important crop yields.

Powdery mildew growing on a leaf (magnified).

Powdery mildew growing on a leaf (high magnification).

Reproduction

Powdery mildew fungi can only reproduce on their living cell host and reproduce both sexually and asexually. Sexual reproduction is via chasmothecia (formerly cleistothecium), a type of ascocarp where the genetic material recombines. Powdery mildew fungi must be adapted to their hosts to be able to infect them. Within each ascocarp are several asci.

Under optimal conditions, ascospores mature and are released to initiate new infections. Conditions necessary for spore maturation differ among species. Asexual reproduction is where the mother fungi and offspring are genetically identical. Powder mildew fungi offspring of wheat and barley species are more successful from asexual reproduction compared to sexual reproduction counterparts.

Vectors of Transmission

Woolly aphids (Eriosomatinae) and other sucking insects are often vectors of transmission for powdery mildew, and other infectious diseases. Typically woolly aphids in sub temperate climates precede and are an indicator of various infections, including Powdery mildew. Aphids penetrate plant surfaces where they often reside and provide a host of potential inoculants through physical, digestive or fecal secretions. Aphids are often an indicator of other potential plant problems.

Management

In an agricultural setting, the pathogen can be controlled using chemical methods, genetic resistance, and careful farming methods.

Conventional Chemical Control

Standard fungicides are an effective way to manage powdery mildew disease on plants. Spray programs of conventional fungicides are advised to begin when powdery mildew symptoms and signs are first noticed. Conventional fungicides should be applied on a regular basis for best results against the disease.

Control is possible with triadimefon and propiconazole. It is also possible with hexaconazole, myclobutanil, and penconazole.

Non-conventional Chemical Control

There are some unconventional chemical control methods that offer alternative modes of action.

The most effective non-conventional methods of chemical control against powdery mildew are milk, natural sulfur (S_8), potassium bicarbonate, metal salts, and oils.

Powdery mildew on a maple leaf as seen under a scanning electron microscope.

Metal salt fungicides should be applied on a regular basis up until harvest of the host. Sulfur must be applied before the disease has emerged since it prevents fungi spores from germinating. Copper sulfate is an effective fungicide allowed in organic farming, but can cause harm to the host plant. Addition of lime hampens this effect.

Neem oil effectively manages powdery mildew on many plants by interfering with the fungus' metabolism and terminating spore production. Sulfur and Fish Oil + Sesame Oil is a mixture effective against powdery mildew.

Milk has long been popular with home gardeners and small-scale organic growers as a treatment for powdery mildew. Milk is diluted with water (typically 1:10) and sprayed on susceptible plants at the first sign of infection, or as a preventative measure, with repeated weekly application often controlling or eliminating the disease. Studies have shown milk's effectiveness as comparable to some conventional fungicides, and better than benomyl and fenarimol at higher concentrations. Milk has proven effective in treating powdery mildew of summer squash, pumpkins, grapes, and roses. The exact mechanism of action is unknown, but one known effect is that ferroglobulin, a protein in whey, produces oxygen radicals when exposed to sunlight, and contact with these radicals is damaging to the fungus.

Dilute sprays containing sodium bicarbonate (baking soda) and vegetable or mineral oils in water are often recommended for controlling powdery mildew, but such mixtures have limited and inconsistent efficacy. While sodium bicarbonate has been shown to reduce to growth of mildews in lab tests, sprays containing only baking soda and water are not effective in controlling fungal diseases on infected plants, and high concentrations of sodium are harmful to plants.

Potassium bicarbonate is an effective fungicide against powdery mildew and apple scab, allowed for use in organic farming.

Another non-conventional chemical treatment involves treating with a solution of calcium silicate. Silicon helps the plant cells defend against fungal attack by degrading haustoria and by producing callose and papilla. With silicon treatment, epidermal cells are less susceptible to powdery mildew of wheat.

Genetic Resistance

Pm3 allel is an effective genetic resistance strategy that protects host species against powdery mildew fungus.

Powdery Mildews of Various Plants

Wheat, Barley and other Cereals

Blumeria graminis f. sp. *tritici*, causes powdery mildew of wheat, whereas f. sp. *hordei* causes powdery mildew of barley.

Legumes

Legumes, such as soybeans, are affected by *Microsphaera diffusa*.

Powdery mildew on Soybean leaves.

Grape

Erysiphe necator (or *Uncinula necator*) causes powdery mildew of grapes.

Powdery mildew of grape.

Onions

The fungus causing powdery mildew of onions is *Leveillula taurica* (also known by its anamorph name, *Oidiopsis taurica*). It also infects the artichoke.

Apples and Pears

Podosphaera leucotricha is a fungus that can cause powdery mildew of apples and pears.

Gourds and Melons

Powdery mildew of cucurbits.

Multiple species of fungus can cause powdery mildew of cucurbits: cucumbers, squashes (including pumpkins), luffas, melons, and watermelons.

Since 1925, commercial *Cucumis melo* (cantaloup and muskmelon) production has been engaged in a biological "arms race" against cucurbit powdery mildew (CPM) caused by the fungus *Podosphaera xanthii*, with new cultivars of melons being developed for resistance to successively arising races of the fungus, identified simply as race 1, race 2, etc. (seven in total by 2004), for races found around the world, and race N1 through N4 for some divergent races native to Japan. Various sub-races have been identified, and given names such as race 2U.S. race 3.5, and race 4.5. A new race S was discovered in 2003, and a specific melon cultivar (*C. melo* var. *acidulus* 'PI 313970') found resistant to it, then used for backcrossing to increase resistance in other cultivars. Such modern selective breeding of plants for phytopathological resistance to particular fungal races involves a great deal of genetic research; this PI 313970 versus race S case involved multi-stage hybridization to propagate a recessive gene, *pm-S* in successive generations, and how this may affect other recessive and codominant genes for resistance to other races of *P. xanthii* "remains to be determined".

A 2004 literature review regarding powdery mildew races that parasitize various cucurbit plants concluded that "race identification is important for basic research and is especially important for the commercial seed industry, which requires accuracy in declaring the type and level of resistance in its products". However, identifying specific races was seen as having little utility in horticulture for choosing specific cultivars, because of the rapidity with which the local pathogen population can change geographically, seasonally, and by host plant.

At least three other Erysiphaceae fungi can cause powdery mildew in cucurbits: The most frequent, after *P. xanthii*, is *Erysiphe cichoracearum*, the former primary causal organism throughout most of the world. *Podosphaera fusca* is another, sometimes considered synonymous with *P. xanthii*. Cucumbers in greenhouse environments have also been reported to be susceptible to *Leveillula taurica*.

Lilacs

Microsphaera syringae is a fungus that can cause powdery mildew in lilac.

Strawberries

Podosphaera aphanis is the cause of powdery mildew in strawberries and other Rosaceae like *Geum rivale* (the water avens).

Tree Leaves

Sawadaea tulasnei is a fungus that causes powdery mildew on tree leaves. This fungus attacks the leaves of the *Acer platanoides* (Norway maple) in North America, Great Britain, and Ireland, *Acer palmatum* (also known as the Japanese maple or smooth Japanese maple).

Oregon Grape

Erysiphe berberidis is a fungus that causes powdery mildew on Oregon grape leaves.

Arabidopsis

Golovinomyces orontii causes powdery mildew on *Arabidopsis* (rockcress) leaves.

Hyperparasites of Powdery Mildew

In the family Sphaeropsidaceae of the Sphaeropsidales fungi, species of the genus *Cicinnobolus* are hyperparasites of powdery mildew.

Ampelomyces quisqualis is an anamorphic fungus that is a hyperparasite of powdery mildews. This parasitism reduces growth and may eventually kill the mildew. Research on biological control of powdery mildews (especially in high-value crops such as grapes) has been ongoing since the 1970s, resulting in the development of fungicides which contain *A. quisqualis* as the active ingredient

Soybean Rust

Soybean rust is a disease that affects soybeans and other legumes. It is caused by two types of fungi, *Phakopsora pachyrhizi*, commonly known as Asian soybean rust and *Phakopsora meibomiae*, commonly known as New World soybean rust. *P. meibomiae* is the weaker pathogen of the two and generally does not cause widespread problems. The disease has been reported across Asia, Australia, Africa, South America and the United States.

Host and Symptoms

Soybean rust is caused by two types of fungi, *Phakopsora pachyrhizi* and *Phakopsora meibomiae*. It affects several important commercial plants, however, most notable for soybeans. Asian

Soybean Rust can infect and reproduce on 90 known plant species, 20 of which are found in the United States, such as, soybeans, dry beans, kidney beans, peas, leguminous forage crops such as trefoil and sweet clover and weeds such as kudzu.

Soybean leaves infected with ASR.

At the early stage of Asian Soybean Rust, it causes yellow mosaic discoloration on the upper surfaces of older foliage. At this stage, it is usually hard to identify since the symptoms are relatively small and poorly defined.

Later as the disease continues to progress, the leaves will turn yellow and there will be lesions mostly on the undersides of the leaves and sometimes on petioles, stems or pods and premature defoliation can also be observed.

Asian Soybean Rust produces two types of lesions. Lesions at the later stage will turn from gray to tan or reddish brown. Mature tan lesion consists of small pustules which surrounded by discolored necrotic areas. Tan spores can be found at the necrotic areas on the underside of the leaf. For Reddish brown lesion, it has larger reddish brown necrotic areas with few pustules and visible spores on the underside of the leaf. A good way to distinguish Asian Soybean Rust from other diseases is to look at the pustules it produces. ASR pustules usually do not have the yellow halo which is related to bacterial pustule. Besides, ASR pustules are raised and can be commonly found on the underside of the leaf which makes it different from the lesions caused by spot diseases.

As one of ASR's most known hosts, soybean plants are susceptible at any stage in the life cycle. However, symptoms are most commonly found during or after flowering. Soybean plants infected by Asian Soybean Rust will result in declining of pod production and fill.

Disease Cycle

Soybean rust is spread by windblown spores and has caused significant crop losses in many soybean-growing regions of the world. Windblown spores can travel for great distances and are released in cycles of seven days to two weeks. It is likely that ASR will survive on vast acreages of naturalized kudzu in the southern U.S. and thereby establish a permanent presence in the continental U.S. It is commonly believed that the disease was carried from Venezuela to the United States by Hurricane Ivan.

Phakopsora pachyrhizi is an obligate parasite, meaning that it must have live, green tissue to survive. For this reason ASR is something that will blow in every year, as cold winters will push it back. It can overwinter in southern states, so long as it has a living host.

ASR overwinters on live host legumes and sporulates the following spring. It cannot survive on dead tissue or crop residues.

Additional hosts can serve as overwintering reservoirs for the pathogen and allow for build-up of inoculum, in those environs free from freezing temperatures. The pathogen is well adapted for long-distance dispersal, because spores can be readily carried long distances by the wind to new, rust-free regions.

Overwintering sites of soybean rust are restricted to areas with very mild winters, such as the gulf coasts of Florida, the very southernmost areas of Texas, or in Mexico. Soybean rust will not survive over the winter in the North Central region because it can't live and reproduce without green living tissue.

Spores of the soybean rust pathogen are transported readily by air currents and can be carried hundreds of miles in a few days. Weather conditions will determine when and where the spores travel from south to north.

Rust spores, called Urediniospores, are able to penetrate the plant cells directly, rather than through natural openings or through wounds in the leaf tissue. Thus infection is relatively quick: about 9 to 10 days from initial infection to the next cycle of spore production.

Rust is a multi-cyclic disease. After the initial infection is established, the infection site can produce spores for 10 to 14 days. Abundant spore production occurs during wet leaf periods (in the form of rain or dew) of at least 8 hours and moderate temperatures of 60 to 80 °F (15.6 to 26.7 °C).

The Process

The infection process starts when urediniospores germinate to produce a single germ tube that grows across the leaf surface, until an appressorium forms. Appressoria form over anticlinal walls or over the center of epidermal cells, but rarely over stomata. Penetration of epidermal cells is by direct penetration through the cuticle by an appressorial peg. When appressoria form over stomata, the hyphae penetrate one of the guard cells rather than entering the leaf through the stomatal opening. This rust and related species are unique in their ability to directly penetrate the epidermis; most rust pathogens enter the leaf through stomatal openings and penetrate cells once inside the leaf. The direct penetration of the epidermal cells and the non-specific induction of appressoria in the infection process of *P. pachyrhizi* may aid in understanding the broad host range of the pathogen and may have consequences in the development of resistant cultivars.

Uredinia can develop 5 to 8 days after infection by urediniospores. The first urediniospores can be produced as early as 9 days after infection, and spore production can continue for up to 3 weeks. Uredinia may develop for up to 4 weeks after a single inoculation, and secondary uredinia will arise on the margins of the initial infections for an additional 8 weeks. Thus, from an initial infection, there could be first generation pustules that maintain sporulation for up to 15 weeks. Even under dry conditions this extended sporulation capacity allows the pathogen to persist and remain a threat. If conditions for re-infection are sporadic throughout the season, significant inoculum potential still remains from the initial infection to reestablish an epidemic. Successful infection is dependent on the availability of moisture on plant surfaces. At least 6 hours of free moisture is needed for infection with maximum infections occurring with 10 to 12 hours of free moisture. Temperatures between 15 and 28 °C are ideal for infection.

Management and Control

Disease control options for ASR are limited. Rust descends in clouds of spores across the countryside. Cultural practices such as row spacing and crop rotations have little effect. Resistant cultivars do not exist. When weather and disease infection conditions are favorable, the occurrence of ASR can be widespread. Thus, remedial control measures—using fungicides as protective sprays—are the only effective disease control method.

Synthetic fungicides are the primary disease control option for protection against Asian soybean rust. The cost of spraying is estimated to be about $15 to $20 per acre; however, two or three sprays may be needed over the course of the growing season. These are significant additional production costs for soybean growers.

Fungicide screening trials to determine disease control efficacy have been field conducted in South America and South Africa. These reports are available on the Web through USDA's Integrated Pest Management Information Centers. These research trials form the basis for fungicidal recommendations in the U.S.

Rust-resistant varieties of soybeans are currently in development by both public universities and private industry.

In some regions, the selection of winter cover crops and forage legumes may be effected, since they can serve as host plants. Resistant soybean varieties are not yet available. However, resistance genes have been identified and host resistance is expected to be an effective, long-term solution for soybean rust. Until resistant commercial varieties are in place, the management of rust depends on judicious use of fungicides.

When untreated, soybean rust, causes yield losses due to premature defoliation, fewer seeds per pod and decreased number of filled pods per plant.

Glomerella Graminicola

Glomerella graminicola ((anamorphic) *Colletotrichum graminicola*) is a fungus in the teleomorphic phase whose anamorphic phase, *Colletotrichum graminicola,* causes anthracnose in many cereal species including maize and wheat. Corn is affected in large numbers in the United States by this fungus, especially certain varieties that have been genetically engineered. These engineered varieties are more susceptible to the teleomorph phase of the fungus. It is not until the fungus moves to the teleomorph phase of the lifecycle and begins to produce fruiting bodies that host plants will begin to exhibit symptoms, often on plants depleted in energy after the stress of pollination. Once the pathogen is in a field, producers can suffer huge economic losses. The disease, corn anthracnose leaf blight, is the most common stalk disease in maize and occurs most frequently in reduced-till or no-till fields. As these practices are widespread, as can be the pathogen.

Anthracnose stalk rot.

Host and Symptoms

C. graminicola is a fungal pathogen that colonizes and infects many turfgrasses, e.g. bluegrass, ryegrass, fescue. In addition to the grasses, C. *graminicola* also infects many grain crops such as barley, wheat, sorghum and corn. The fungus can infect many different parts of the corn plant, typically the kernels, tassels, roots, leaves, stalk and husks. The most common area of infection is the stalk. C. *graminicola* produces three major symptom types: leaf blight, stalk rot and top die-back. The leaf blight is characterized by round yellowing water soaked lesions on the leaves. These lesions usually occur early in the season and are how this pathogen is distinguished from other diseases. Top die-back is the necrosis of the top leaves and stalk of the corn. This occurs around the same time as grain formation. The stalk rot phase becomes prominent during the late reproductive stages of the corn life cycle. It is characterized by blackening of the pith tissue in the stalk and also of the rind, beginning at the nodes closest to the soil. Along with these symptoms, seedling blight and post emergence damping off are also found.

Identification

Stromata

- 70-300 μm in diameter,
- Bear prominent, dark, septate spines (setae) up to 100 μm long.

Conidia

- Developing at the base of the spines,
- Hyaline to pale yellow, unicellular, sickle-shaped, falcate to fusiform, tapered toward both ends,
- 3-5 x 19-29 μm.

Phialides

- Unicellular, hylanine and cylindrical,
- 4-8 x 8-20 μm.

Growth on PDA

- Gray and feltlike,
- Conidia and appressoria are numerous when culture are well aerated, and sclerotia sometimes occur.
- Appressoria are diagnostic: they are tawny brown, irregular-shaped in edge, prominent, and terminal on thickened hyphae.

Disease Cycle

In the spring, fruiting structures (acervuli) form from corn residue and produce spores (conidia) that are dispersed by wind blown raindrops and splashing. Conidial spores infect young plants through the epidermis or stomata. Anthracnose develops rapidly in cloudy, overcast conditions with high temperatures and humidity. In optimal environmental conditions, conidia can germinate in as little as 6–8 hours in 100% humidity. Initial necrotic spots or lesions can be seen within 72 hours after infection by conidia. Lower leaves that develop lesions provide conidial spores and cause secondary infections on the upper leaves and stalk. Vascular infections primarily occur from wounds caused by stalk-boring insects, such as the larvae of the European corn borer, allowing for conidia to infect and colonize the xylem. From this, anthracnose top die back (vascular wilt) or stalk rot can occur. In the fall, *C. graminicola* survives as a saprophyte on corn leaf residue. The pathogen can also overwinter on corn stalks as conidia in an extracellular secretion. The secretion prevents conidia from desiccating and protects them from unfavorable environmental conditions. Overwintering on corn residue serves as a vital source of primary inoculum for the leaf blight phase in the spring. The cycle will start all over again when susceptible corn seedlings emerge from the ground in the spring.

Disease Management

Since C. graminicola is found to survive on corn residue, specifically on the soil surface; one of the most effective methods of control is a one-year minimum of crop rotation to reduce anthracnose leaf blight. A study in 2009 showed more severe symptoms of leaf blight due to C. graminicola when grown on fields previously used for corn in comparison to fields previously used for soybean. Other management methods include the use of hybrid selections and tillage systems. Keeping in mind, hybrid selection may be resistant to leaf blight but they are not necessary resistant to other fungal diseases such as stalk rot. Tillage systems that are able to fully bury corn residue deep underground along with one year crop rotation will reduce the source of inoculum greatly. More work is still needed in order to determine the influence of buried and surface corn residues as a source of inoculum for corn anthracnose.

Pathogenesis

Once conidia germinate on corn leaves, a germ tube differentiates and develops into an appresoria and allows *C. graminicola* to penetrate epidermal cells. Germination and appressorium formation occur best in the temperature range (15-30 °C) Penetration occurs in a much narrower temperature range (25-30 °C). In order to penetrate the cell wall, the fungus first pumps melanin into the walls of the appressorium to create turgor pressure in the appressorium. The melanin allows water into the appressorium cell but nothing out. This builds up an incredible amount of turgor pressure which the fungus then uses to push a hyphae through the corn cell wall. This is called the penetration peg. The penetration peg then grows, extends through the cell extracting nutrients and the host cell wall dies. Hyphae migrate from epidermal cells to mesophyll cells. As a defense response, the cells produce papillae to prevent cell entry but is typically not seen successful. It is believed C. graminicola has a biotrophic phase because the plasma membrane of the epidermal cells is not immediately penetrated after invasion into the epidermal cell wall. Between 48–72 hours after infection, *C. graminicola* shifted from biotrophic growth to nectrotrophy (lesions appear). This is when secondary hyphae invade cell walls and intercellular spaces.

Black Sigatoka

Black Sigatoka is a leaf-spot disease of banana plants caused by the ascomycete fungus *Mycosphaerella fijiensis* (Morelet). Also known as black leaf streak, it was discovered in 1963 and named for its similarities with yellow Sigatoka, which is caused by *Mycosphaerella musicola* (Mulder), which was itself named after the Sigatoka Valley in Fiji, where an outbreak of this disease reached epidemic proportions from 1912 to 1923.

The Sigatoka disease complex is a cluster of three closely related fungi: yellow Sigatoka (*Pseudocercospora musae*), eumusae leaf spot (*Ps. eumusae*), and black Sigatoka (*Ps. fijiensis*).

Plants with leaves damaged by the disease may have up to 50% lower yield of fruit, and control can take up to 50 sprays a year.

M. fijiensis reproduces both sexually and asexually, and both conidia and ascospores are important in its dispersal. The conidia are mainly waterborne for short distances, while ascospores are carried by wind to more remote places (the distances being limited by their susceptibility to ultraviolet light). Over 60 distinct strains with different pathogenic potentials have been isolated. To better understand the mechanisms of its variability, projects to understand the genetic diversity of *M. fijiensis* have been initiated.

When spores of M. fijiensis are deposited on a susceptible banana leaf, they germinate within three hours if the humidity is high or a film of water is present. The optimal temperature for germination of the conidia is 27 °C (81 °F). The germ tube grows epiphytically over the epidermis for two to three days before penetrating the leaf through a stoma. Once inside the leaf, the invasive hypha forms a vesicle and fine hyphae grow through the mesophyll layers into an air chamber. More hyphae then grow into the palisade tissue and continue on into other air chambers, eventually emerging through stomata in the streak that has developed. Further epiphytic growth occurs before the re-entry of the hypha into the leaf through another stoma repeats the process. The optimal conditions for M. fijiensis as compared with M. musicola are a higher temperatures and higher relative humidity, and the whole disease cycle is much faster in M. fijiensis.

Symptoms

Black Sigatoka is also known as black leaf streak. The pathogen Mycosphaerella fijiensis causes streaks that run parallel to the leaves. It is an ascomycete fungus that affects banana trees specifically in tropical climates; including Asia, West Africa, China, and South America. Tropical weather is the preferred climate for banana cultivation, but it is also the environment where the pathogen thrives: hot and humid, with plenty of rainfall to aid in dispersal. The optimal environment of the pathogen is similar to that of the banana tree. The fungus infects mature banana leaves and will continue to cause infection without proper control.

Black Sigatoka lesions on mature banana leaf.

In the early stages of the infection of the plant, the lesions have a rusty brown appearance and appear to be faint, paint-like specks on the leaves. They become more visible on the undersides of the banana leaf as the lesions and leaves grow. The spots on the undersides of leaf are the fungus itself. The sign of the pathogen consists of the ascocarp which holds the ascospores used for dissemination to infect healthy new plants when the environment is conducive. The pathogen then survives on dead plant tissue as mycelium. The dimensions of the lesions are characteristically 20 x 2mm with a well defined wall surrounding it. After further development, they become darker, sink into the leaf, and turn into depressions. The depressions themselves and the chlorosis surrounding them are the visible symptoms of the plant pathogen. They eventually will merge, causing the rapid decline of plant morphological and physiological function. Leaves with large infectious lesions will start to degrade and collapse because the leaf spots interrupt the plant's ability to perform photosynthesis, leading to the ultimate death of the plant.

The yellow leaf streak pathogen is in the same genus as that of black leaf streak. Yellow leaf streak shows smaller, yellow-green lesions that appear on top of the leaves.

Management

There are several ways to control black Sigatoka, either by cultural and chemical means or by genetic engineering. Cultural control includes the destruction of leaves that have been infected with *M. fijiensis*. This will help reduce the initial (ascospores) and secondary (conidia) spread of inoculum of new plant leaves and interrupt the pathogen's polycyclic disease cycle. Another way of reducing primary/secondary inoculums is via efficient drainage and irrigation. Keeping the environment around the plants at low humidity helps keep the ascospores/conidia produced by the pathogen from being dispersed in the water draining towards other healthy, susceptible hosts. Other techniques include planting the banana trees more than 1,000 meters above sea level and practicing multi-cropping, mixing banana with other trees or vegetation.

One form of chemical control is applying fungicides. This is a preemptive control used on banana trees in order to protect them from primary inoculum. The fungicide does not kill the pathogen itself, but works on the pre-necrotic spots on the leaves, stopping the secondary spores from inoculating new, healthy plant tissue. The best time to apply this protective fungicide is in the beginning of the season in order to stop any initial infection. The class of fungicides widely used to control black leaf streak is the triazoles. These are demethylation inhibitors and should be rotated with compounds having other modes of action to slow the development of resistance. Leaves that have already been infected must be removed mechanically to save the rest of the tree. Research has shown that there may be fungicide resistance developing for M. fijiensis. It has been observed that following the intensive application of chemicals, the fungus persisted and spread. The same observations were found in fields with no chemical interference; the belief now being that the untreated fields are "breeding grounds for the development of resistant strains". Research today shows continuous action towards reinventing banana breeding programs. However, some cultivars of bananas are resistant to the disease. Research is done to improve productivity and fruit properties of these cultivars. A genetically modified banana variety made more resistant to the fungus was developed and was field tested in Uganda in the late 2000s. Furthermore, the search for genetic resistance shows promise with the discovery of a protein that can produce a hypersensitive response to control M. fijiensis that is being introduced into banana trees. This may lead to the identification of a resistance gene that could be transferred to banana trees.

BACTERIAL DISEASES

Aster Yellows

Witches'-broom of an infected carrot.

Aster yellows is a chronic, systemic plant disease caused by a bacterium-like organism called a phytoplasma. The aster yellows phytoplasma (AYP) affects 300 species in 38 families of broad-leaf herbaceous plants, primarily in the aster family, as well as important cereal crops such as wheat and barley. Symptoms are variable and can include phyllody, virescence, chlorosis, stunting, and sterility of flowers. The aster leafhopper vector, *Macrosteles quadrilineatus*, moves the aster yellows phytoplasma from plant to plant. Its economic burden is primarily felt in the carrot (*Daucus carota* ssp. *sativus*) crop industry, as well as the nursery industry. No cure is known for plants infected with aster yellows. Infected plants should be removed immediately to limit the continued spread of the phytoplasma to other susceptible plants. However, in agricultural settings such as carrot fields, some application of chemical insecticides has proven to minimize the rate of infection by killing the vector.

Hosts and Symptoms

Aster yellows affects a long list of plant species including native plants, ornamentals, weeds, and vegetables crops. The largest family affected is the Asteraceae, and ornamental plants commonly infected are asters, marigolds, coreopsis, and purple coneflower. Regarding vegetable crops, onion, lettuce, celery, and carrot are affected with the latter suffering the greatest losses.

The range of characteristic symptoms varies by the phytoplasma strain, timing of infection, plant species, temperature, age, and size of the plant. The symptoms can be mistaken for herbicide damage. They include vein clearing until the entire leaf becomes chlorotic, stunting, deformation, virescence (greening of flowers), phyllody (development of leaf-like flower petals), reddening of foliage, reduced root system, and sterility. Aster yellows does not typically kill perennial host plants.

Characteristic symptoms specific to the carrot include initial vein clearing and chlorosis, followed by production of many adventitious shoots, with the tops looking like a witches'-broom. The internodes of such shoots are short as are the leaf petioles. Young leaves are smaller and dry up while the petioles of older leaves twist and break off. Any remaining older leaves turn bronze or red late in the season. Floral parts are deformed and roots are smaller, abnormally shaped, and have woolly secondary roots. The carrot roots are predisposed to soft rots in the field and storage and taste unpleasant to the consumer.

Aster leafhopper.

Disease Cycle

The aster yellows disease is caused by the aster yellows phytoplasma (AYP) which is a phloem-limited, bacterium-like organism and is vectored by the aster leafhopper, *Macrosteles quadrilineatus*, a phloem-feeding insect of the order Hemiptera.

Phytoplasmas are small (0.5-1 µm in diameter) prokaryotes that reproduce by division or budding in the phloem sieve cells of the host plants, as well as the bodies of their leafhopper vectors. Currently, AYP cannot be cultured in cell-free media, making detailed study somewhat more challenging. AYP has the ability to increase the fecundity and lifespan of their insect vector, thus enhancing the ability of the host to transfer AYP from plant to plant. AYP survives in perennial weeds, ornamentals, and vegetables. Some examples of weed host plants are thistle, wild carrot, dandelion, field daisy, black-eyed Susan, and wide-leafed plaintain.

The vector leafhopper feeds on the phloem of aster yellows-infected plants by inserting its straw-like mouthpart, a stylet, into the cell and extracting it. Once the phytoplasma is acquired, an incubation period follows in which it multiplies within the leafhopper and then moves to the salivary glands. At this point, the phytoplasma can be transmitted to a new host through the saliva as the leafhopper feeds. Within 8–24 hours after inoculation, the phytoplasma moves out of the leaf into the host plant phloem. Cells adjacent to the phloem enlarge and die while surviving cells begin to divide, but soon die, too. Surrounding cells in the region of the necrotic area begin to divide and enlarge, producing abnormal sieve elements, while the phloem elements within the necrotic areas degenerate and collapse. Infected plants usually show symptoms after 8–9 days at 25 °C and 18 days at 20 °C, with no symptoms developing at 10 °C.

Environment

Hardly any conditions directly affect the development of aster yellows, but a few indirect factors strongly influence the rate of transmission by the leafhopper. Conditions that favor movement and spread of the leafhopper and encourage feeding assist in the spread of the phytoplasma.

Transcontinental migration begins in the spring when the prevailing winds and jet streams help carry the leafhoppers from their overwintering sites in the South to the Midwest. Upon arrival in the Midwest, they begin feeding. The leafhopper may have migrated into the region already carrying the phytoplasma, which it could have acquired from infected plants along the migration or while still in the South. The leafhopper could have also arrived not yet carrying the phytoplasma. If this is the case, it could feed on perennial weeds that are infected to acquire AYP. Weather conditions greatly influence leafhopper flight because they are poor flyers. Temperatures below 15 °C or rainfall temporarily halt their migration and delay the time of infection. The leafhoppers then feed all summer until they migrate back to their overwintering sites in the fall.

Weather conditions of the region also greatly influence leafhopper feeding patterns. If conditions are hot and dry plants do not appear as lush and nutrient-rich to the phloem-feeding leafhopper, whereas seasons with abundant rainfall allow the plants to have much more lush growth. This means that hot and dry conditions are less conducive to the spread of aster yellows than times of abundant rainfall.

In the Western United States, no migration of the vector leafhoppers occurs. This allows for transmission of the phytoplasma year round.

Management

Aster yellows phytoplasma is a difficult pathogen to control. Currently, no cure for aster yellows is known. Infected plants and weeds should be removed to eliminate that source of the phytoplasma and minimize spread. Unfortunately, this is the only control method that home gardeners have available.

On an agricultural level, speaking specifically about carrots, some methods can be used to manage the leafhopper populations in an attempt to control AYP spread. The aster yellows index (AYI) can be used to determine when to apply chemical controls. The AYI equals the percentage of leafhopper population containing AYP multiplied by the number of leafhoppers present per 100 sweeps. The resulting number can determine when to apply insecticides based on how susceptible the crop or cultivar is to leafhopper feeding. For highly susceptible crops or cultivars, an AYI of 50 indicates the need for application, while for intermediate crops or cultivars the AYI is 75 and for crops or cultivars relatively resistant to economically harmful symptoms the AYI is 100.

Bacterial Wilt

Bacterial wilt is a cucurbit disease caused by the pathogen *Erwinia tracheiphila,* a Gram-negative bacterium in the family Enterobacteriaceae. Cucumber and muskmelon plants are most susceptible, but squash, pumpkins, and gourds may also become infected.

Disease Transmission

E. tracheiphila is spread between plants by two species of insect vectors, Striped cucumber beetles (*Acalymma Vittatum*) and spotted cucumber beetles (*Diabrotica undecimpunctata*). The beetles acquire *E. tracheiphila* by feeding on infected plants, then carry the bacteria in their digestive tracts. The disease may be spread to susceptible plants through feeding wounds, by way of infected mouthparts or frass. The bacteria is capable of overwintering in the gut of its insect vectors.

Symptoms and Diagnosis

Bacterial wilt is a disease of the vascular tissue. When a plant is infected, *E. tracheiphila* multiplies within the xylem, eventually causing mechanical blockage of the water transport system. The first sign of infection, which appears about five days after acquisition, is the wilting of individual leaves on a single stem. However, the disease will soon spread down the runner and then infect the whole plant, causing it to shrivel and die. There is a diagnostic test for bacterial wilt that can be done in the field. The presence of the *E. tracheiphila* causes the sap to become a milky color and acquire a sticky consistency. If the stem is cut near the crown and the ends are slowly pulled apart, the sap should form a viscous string.

Treatment and Prevention

Once a plant is infected, there is no way of stopping the spread of the disease. Some cucurbit cultivars are less susceptible than others, so it is beneficial to plant these cultivars. However, since wilt-resistant plants have not yet been developed, the most effective way to prevent the disease is to keep beetle populations at a minimum. While various methods of beetle control have been tested, the most effective preventative measure is to keep beetle populations as low as possible through careful field monitoring and insecticide sprays.

Bacterial Spot

Bacterial spot is caused by two species of bacteria, Xanthomonas axonopodis pv. vesicatoria and X. vesicatoria. Bacterial spot can be damaging to both pepper and tomato in the High Plains during warm (75 to 86 °F), humid, rainy weather. Bacterial spot does not affect eggplant. Bacteria are introduced onto plants by planting contaminated seed and transplants, splashing rain or irrigation water, aerosols, or on contaminated equipment. The bacteria multiply on leaves to form large populations before penetrating through natural openings or wounds created by wind-blown sand, insect feeding, or mechanical injury. The bacterial spot pathogens survive between susceptible crops in and on weed, volunteer plants, infested crop debris, culls, and contaminated seed and transplants.

Plant Response and Damage

The bacterial spot pathogens can infect all aboveground plant parts. Disease symptoms begin as small, brown, water-soaked lesions, which turn brown with necrotic centers. Leaf lesions are generally sunken on the upper surface, but are raised on lower surfaces. Lesions are rarely larger than 0.12 inch, but when conditions are favorable for disease, lesions coalesce and from large blighted areas. Infected leaves turn yellow and drop prematurely. Fruit lesions begin as small (0.04 inch) circular green spots, but turn brown and become cracked and roughened with age. Bacterial spot reduces both yield and fruit quality. Infected fruit is generally unmarketable.

Management Approaches

Biological Control

Bacteriophage, viruses that attacked bacteria, control bacterial spot but must be applied at dusk at least twice weekly to be effective. Nonpathogenic Xanthomonas spp. provide some control of bacterial spot.

Cultural Control

Plant only high quality seed and transplants free from the bacterial spot pathogens. Hot water treatments can reduce seed contamination, but may reduce germination. Practice a three-year or longer crop rotation between susceptible crops; do not plant tomato and pepper consecutively in a field. Eliminate weeds, volunteers, crop debris, and cull piles that can serve as inoculum sources. Avoid reuse of irrigation tail water and overhead irrigation if possible. Resistant varieties are available, but should be chosen carefully to match the most prevalent pathogenic races of the pathogen present. Eleven pathogenic races are known to occur; no single resistance gene will provide resistance to all pathogenic races.

Chemical Control

Resistance to copper bactericides and streptomycin are widespread in the bacterial spot pathogens. Tank-mixing copper bactericides with EBDC fungicides such as maneb can provide some suppression of copper-tolerant strains of X. axonopodis pv. vesicatoria. The plant activator Actigard can provide effective control of both copper sensitive and tolerant strains of the pathogen, but can reduce yields in the absence of disease. Chemical controls are most effective when combined with as many cultural and biological controls as possible.

Bacterial Blight

Common bacterial blight is caused by the pathogen Xanthomonas axonopodis pv. phaseoli. The pathogen can be found on the leaves of many plants, but only causes disease on a dry bean. Infection occurs when bacterial cells are deposited onto leaves by splashing water, aerosols, or from contaminated seed, and multiply to form large populations. The bacteria gain entry into plants through natural openings and wounds. Infection occurs most readily during warm (greater than 85 °F), wet weather, especially hard, wind-driven rains. Bacteria are disseminated within and among fields by splashing water, aerosols, and on contaminated equipment and workers. The pathogen survives between susceptible hosts in and on weeds, infested crop debris and contaminated seed.

Plant Response and Damage

Common bacterial blight symptoms first appear as flaccid, small water-soaked spots on the underside of leaflets. These spots enlarge and merge, becoming dried and brown. A narrow, bright lemon-yellow border of tissue encircles the lesion. Infected pods develop circular water-soaked spots, and yellow masses of bacteria may appear at their center. Later, the spots dry and become reddish-brown, sunken lesions. Early pod infection causes shriveled seeds, and the bacteria may cause yellow to orange discoloration under the seed coat of infected seeds. A stem girdling or joint rot occurs above the cotyledonary node of plants grown from infected seeds. Yield losses can range from 20 to 40% during certain year. The disease also reduces seed size and quality.

Management Approaches

Biological Control

No biological control strategies have been commercialized for common bacterial blight.

Cultural Control

Plant certified seed of recommended varieties less susceptible to common bacterial blight. Avoid overhead irrigation and reuse of irrigation water where possible. Avoid working in fields when plants are wet. Promptly and thoroughly incorporate infested bean debris into the soil after harvest, and rotate beans with non-host crops such as small grains for at least two years. Practice strict sanitation of weeds and volunteer beans early in the following season.

Chemical Control

Antibiotic seed treatment and preventative bactericide applications can reduce spread of the common bacterial blight pathogen, but chemical controls are most effective when integrated with sound cultural practices.

Fire Blight

Fire blight, also written fireblight, is a contagious disease affecting apples, pears, and some other members of the family Rosaceae. It is a serious concern to apple and pear producers. Under optimal conditions, it can destroy an entire orchard in a single growing season.

The causal pathogen is *Erwinia amylovora*, a Gram-negative bacterium in the family Enterobacteriaceae. Pears are the most susceptible, but apples, loquat, crabapples, quinces, hawthorn, cotoneaster, *Pyracantha*, raspberry and some other rosaceous plants are also vulnerable. The disease is believed to be indigenous to North America, from where it spread to most of the rest of the world.

Fire blight is not believed to be present in Australia though it might possibly exist there. It has been a major reason for a long-standing embargo on the importation of New Zealand apples to Australia. Japan was likewise believed to be without the disease, but it was discovered in pears grown in northern Japan. Japanese authorities are, however, still denying its existence, and the Japanese scientist who discovered it is believed to have committed suicide after his name was leaked to affected farmers. In Europe it is listed as a quarantine disease, and has been spreading along Hawthorn (*Crataegus*) hedges planted alongside railways, motorways and main roads.

Symptoms

Fire blight on a pear tree caused by *Erwinia amylovora*.

Tissues affected by the symptoms of *Erwinia amylovora* include blossoms, fruits, shoots, and branches of apple (Pomoideae), pear, and many other rosaceous plants. All symptoms are above ground and are typically easy to recognize. Symptoms on blossoms include water soaking of the floral receptacle, ovary, and peduncles. This results in a dull, gray-green appearance at 1–2 weeks after petal fall, and eventually tissues will shrivel and turn black. The base of the blossom and young fruit show similar symptoms as infection spreads. Opaque white- or amber-colored droplets of bacterial ooze can be seen on the infected tissue when the environment is high in humidity. Shoots show similar symptoms but develop much more rapidly. A "Shepherd's Crook" can be seen when the tip of the shoot wilts, and diseased shoot leaves typically have blackening along the mid-vein and then die. In number, diseased shoots give the tree a blighted appearance. Initial infection of blossoms and shoots can spread to larger tree limbs. Branches will darken and become water soaked. Advanced infection develops cracks in bark and a sunken surface. Wood under the bark will become streaked with black discoloration. Immature fruit forms water-soaked lesions and later turned black. Bacterial ooze can be found on these lesions. Severe infections result in fruit turning entirely black and shriveling. A primary inoculum of this disease is typically from cankers formed the season before. The factors that determine whether or not cankers become active are not well known, but it is thought that cankers found on larger tree limbs are more likely to become active. It is also thought that age may be a factor.

Dissemination

Gala apple branch with scorched leaves after a severe fire blight infection.

Honeybees and other insects, birds, rain and wind can transmit the bacterium to susceptible tissue. Injured tissue is also highly susceptible to infection, including punctures and tears caused by plant-sucking or biting insects. Hailstorms can infect an entire orchard in a few minutes, and growers do not wait until symptoms appear, normally beginning control measures within a few hours.

Once deposited, the bacterium enters the plant through open stomata and causes blackened, necrotic lesions, which may also produce a viscous exudate. This bacteria-laden exudate can be distributed to other parts of the same plant or to susceptible areas of different plants by rain, birds or insects, causing secondary infections. The disease spreads most quickly during hot, wet weather and is dormant in the winter when temperatures drop. Infected plant tissue contains viable bacteria, however, and will resume production of exudate upon the return of warm weather in the following spring. This exudate is then the source for new rounds of primary infections.

The pathogen spreads through the tree from the point of infection via the plant's vascular system, eventually reaching the roots and/or graft junction of the plant. Once the plant's roots are affected, the death of the plant often results. Over pruning and over fertilization (especially with nitrogen) can lead to watersprout and other midsummer growth that leave the tree more susceptible.

E. amylovora typically makes its entry into its host xylem or cortical parenchyma. It can also enter through stomata, lenticles and hydathodes. It is dispersed by rain and or insects naturally, but this mode of dispersal is very ineffective and can only be effective for local transmission of the pathogen. Aerosols are also suspected in playing a role in its transmission due to the detection of *E. amylovora* in Mediterranean regions. In composition the pathogen is composed of short rods with rounded ends made motile by many peritrichous flagellae. *E. amylovara* is a gram negative bacterium.

Management

Spraying plants with streptomycin or injecting plants with oxytetracycline can prevent new infections. The widespread use of streptomycin spray has led to antibiotic resistance in some areas, such as California and Washington. Certain biological controls consisting of beneficial bacteria or yeast can also prevent fire blight from infecting new trees. The only effective treatment for plants already infected is to prune off the affected branches and remove them from the area. Plants or trees should be inspected routinely for the appearance of new infections. The rest of the plant can be saved if the blighted wood is removed before the infection spreads to the roots. There is no known cure; prevention is the key.

E. amylovora needs to be destroyed externally, before it enters the cell. This is simply because once it enters the host, it spreads during the endophytic phase of pathogenesis. Once this happens external control methods become ineffective. The ideal control method is to apply copper and antibiotics to the plant externally. This is the only effective method and it is indeed preventative. Currently it has been noted that *E. amylovora* has developed a resistance to the antibiotic streptomycin, as do most bacteria due to their flexible ability to transfer preferential genes promoting resistance to certain antibiotics horizontally from species not even similar to it as all bacteria can.

Phytosanitary measures have been employed as the best sanitary measures against *E. amylovora* dispersal. High risk countries are encouraged not to import plants susceptible to the pathogen into their territory because, once the bacteria become established in an area it is nearly impossible to eradicate the disease. Nurseries and orchards in such regions are placed on strict phytosanitary surveillance measures and well-monitored. Imported and infected crops are destroyed as soon as they are noticed since the bacteria spreads very rapidly and eradication methods are usually costly and inefficient.

Pathogenesis

Pathogenicity depends on many different factors such as the production of the siderophore desferrioxamine, metalloproteases, plasmids, and histone-like proteins. However, some essential factors of pathogenicity are variations in the synthesis of extracellular polysaccharides (EPS) and the mechanism of type III secretion system and its associated proteins. EPS helps bacterial pathogens avoid plant defenses, "clog" the host's vascular system, protect bacteria against desiccation and attach to both surfaces and one another. One EPS is amylovoran, a polymer of pentasaccharide repeating units. If a strain of *E. amylovora* can not produce amylovoran it is not pathogenic and can not spread in plants. Levan is another EPS, and a lack of it will slow development of symptoms. Type III secretion systems are used for exporting and delivering effector proteins into the cytosol of host plants. This system mainly consists of Hrc proteins. Motility is another major virulence factor. Since *E. amylovora* is not an obligate biotroph, it is able to survive outside the host which allows it to spread in many ways such as rain.

Blackleg

Blackleg of Potato complete plant wilt in field. These plants can sometimes be lost in the canopy.

Blackleg is a plant disease of potato caused by pectolytic bacteria that can result in stunting, wilting, chlorosis of leaves, necrosis of several tissues, a decline in yield, and at times the death of the potato plant. The term blackleg originates from the typical blackening and decay of the lower stem portion, or leg, of the plant.

Blackleg in potatoes is most commonly caused by *Pectobacterium atrosepticum*, a gram-negative, nonsporulating, facultative anaerobe that is also associated with soft rot of potatoes. While other bacterial species such as *Pectobacterium carotovorum* and *Dickeya dadantii* can exhibit symptoms similar to blackleg of potato, these pathogens exhibit broader host ranges, are present in different climates, and typically are more associated with soft rot diseases.

Symptoms and Signs

Stem discoloration and darkening.

Wilting caused by blackleg.

Early blackleg symptoms develop in the growing season soon after the plants emerge. They are characterized by stunted, yellowish foliage that has a stiff, upright habit. The lower part of the below ground stem of such plants is dark brown to black in color and extensively decayed. When infected, the pith region of the stem is particularly susceptible to decay and may extend upward in the stem far beyond the tissue with externally visible symptoms. Young plants affected by blackleg are particularly susceptible, typically dying after a halt in development.

Blackleg symptoms may develop in more mature plants during the later part of the growing season, and are distinguished from those that develop earlier in the season. Blackleg appears as a black discoloration of previously healthy stems, accompanied by a rapid wilting, and sometimes yellowing, of the leaves. Starting below ground, black discoloration moves up the stem, often until the entire stem is black and wilted. However, in some cases of early disease development, mature stems may turn yellow and wilt even before black decay is evident. However, after the entire stem exhibits disease symptoms, the wilted plant can be lost from view in the healthy potato plant canopy.

Disease Cycle

1. A contaminated tuber can infect growing stems, or move into the vascular bundles of mature stems.

2. Infected stems can be symptomatic or asymptomatic, depending on environmental conditions, although the disease will remain and spread to other tubers on the same plant through the stolons.

3. In the field or during storage, they can contaminate and infect healthy tubers through wounds introduced during harvesting or through lenticels, and may also be spread through insects, wind, and rain. An important insect vector is the seed corn maggot (*Hylemya platura* and *H. florilega*), which spread the bacteria from diseased to healthy tissues. The bacteria are carried in the intestinal tracts of these insects, which spread the pathogen to healthy tissue by feeding on cut surfaces of healthy seed tissue. Another insect vector is the fruit fly (*Drosophila melanogaster*).

4. The pathogen will often survive in the infected tubers until the following planting season.

Environment and Biology

The pathogen *P. atrosepticum* thrives in moist, cool conditions, typically causing symptoms at temperatures below 25 °C. It is vulnerable to temperatures above 36 °C and dry conditions, and thus survives best in potato tuber tissues, although it is known to survive in other plant tissues. Unlike other pectolytic bacteria, evidence shows that *P. atrosepticum* does not survive well in soil outside its host tissue.

Disease symptoms are not necessarily uniformly exhibited from both shoots originating from a single tuber or in a field infested with *P. atrosepticum*. Additionally, presence of *P. atrosepticum* in the soil is not necessarily associated with disease symptoms. This is partly explained by the narrow environmental conditions needed for pathogenicity, although new findings in research are showing strong evidence of density dependent quorum sensing signals used by *P. atrosepticum* in exhibiting virulence.

Management

Cultural

Blackleg of potato has been successfully managed primarily using cultural techniques. These techniques generally rely on sterile propagation techniques, using knowledge *P. atrosepticum's* narrow environmental range to control planting timing, removing infected tissues and plants during the growing season, reducing tuber harvest damage, and proper storage.

Sterile Propagation

Given that tubers are the primary mechanism by which *P. atrosepticum* survives and spreads, clean seed potato stocks established using tissue cultures have been very successful in breaking the cycle of carrying disease forward from year to year. Buildup of tuber contamination is limited by reducing the number of field generations of these seed potatoes to 5 to 7 years. Some methods of sterile propagation include planting only healthy, whole seed potatoes. If healthy seed potatoes are to be cut, they should be first warmed to 12-15 °C, cut, stored for 2 days at 12-15 °C in a humid environment with good air flow. This warming and storing period ensures proper suberization of the tissue, which forms a barrier from *P. atrosepticum* infestation.

Planting Conditions

Given that *P. atrosepticum* thrives in cool, moist conditions, planting seed potatoes in well-drained soil after soil temperatures have increased well above 10 °C is very important to halting the onset of the disease early in the plant life cycle, when the plant is more susceptible to the worst effects of the disease.

Nutrition

Increasing application of nitrogen or complete fertilizers have shown reduced incidence of blackleg stem infection.

Growing Season

Although there is a risk of spreading the disease pathogen through injury of healthy plants, if proper techniques are followed, rogueing out all parts of the blackleg-diseased plants can be a useful way to reduce soil inoculum.

At Harvest and During Storage

Given that *P. atrosepticum* survives best in the tubers and additionally contributes to soft rot, it is critically important to reduce spread of the pathogen by removing tubers exhibiting soft rot decay before they are spread over grading lines and bin pilers for storage. Reducing post-harvest wounding is also important, especially for seed potatoes. Additionally, it is critically important to keep the potatoes at a low temperature with adequate aeration and humidity control in order to minimize development of the pathogen in infested stocks.

Biocontrol and Plant Resistance

New research on *P. atrosepticum* virulence pathways has elucidated the use of quorum sensing molecules to exhibit pathogenicity. These pathways include the control of the production of plant cell wall degrading enzymes in addition to other virulence factors. Research indicating the role of other soil microbes in degrading *P. atrosepticum* quorum sensing communication molecules provides the possibility for safe and effective control of the disease.

Plant defense mechanism studies on *P. atrosepticum*, used to better understand disease resistance, have focused more on the soft-rot symptoms that can sometimes be associated with *P. atrosepticum*. However, research is successfully identifying the quantity and type of plant resistance molecules that are produced in response to pathogen associated molecular patterns (PAMPs), and their effects on the activity and virulence of pathogens such as *P. atrosepticum*.

Citrus Canker

Citrus canker is a disease affecting *Citrus* species caused by the bacterium *Xanthomonas axonopodis*. Infection causes lesions on the leaves, stems, and fruit of citrus trees, including lime, oranges, and grapefruit. While not harmful to humans, canker significantly affects the vitality of citrus trees, causing leaves and fruit to drop prematurely; a fruit infected with canker is safe to eat, but too unsightly to be sold.

The disease, which is believed to have originated in Southeast Asia, is extremely persistent when it becomes established in an area. Citrus groves have been destroyed in attempts to eradicate the disease. Brazil and the United States are currently suffering from canker outbreaks.

Biology

Xanthomonas axonopodis is a rod-shaped Gram-negative bacterium with polar flagella. The bacterium has a genome length around 5 megabase pairs. A number of types of citrus canker diseases are caused by different pathovars and variants of the bacterium:

- The Asiatic type of canker (canker A), X. axonopodis pv. citri, caused by a group of strains originally found in Asia, is the most widespread and severe form of the disease.

- Cancrosis B, caused by a group of X. axonopodis pv. aurantifolii strains originally found in South America is a disease of lemons, key lime, bitter orange, and pomelo.
- Cancrosis C, also caused by strains within X. axonopodis pv. aurantifolii, only infects key lime and bitter orange.
- A* strains, discovered in Oman, Saudi Arabia, Iran, and India, only infect key lime.

Pathology

Plants infected with citrus canker have characteristic lesions on leaves, stems, and fruit with raised, brown, water-soaked margins, usually with a yellow halo or ring effect around the lesion. Older lesions have a corky appearance, still in many cases retaining the halo effect. The bacterium propagates in lesions in leaves, stems, and fruit. The lesions ooze bacterial cells that, when dispersed by windblown rain, can spread to other plants in the area. Infection may spread further by hurricanes. The disease can also be spread by contaminated equipment, and by transport of infected or apparently healthy plants. Due to latency of the disease, a plant may appear to be healthy, but actually be infected.

Citrus canker bacteria can enter through a plant's stomata or through wounds on leaves or other green parts. In most cases, younger leaves are considered to be the most susceptible. Also, damage caused by citrus leaf miner larvae (*Phyllocnistis citrella*) can be sites for infection to occur. Within a controlled laboratory setting, symptoms can appear in 14 days following inoculation into a susceptible host. In the field environment, the time for symptoms to appear and be clearly discernible from other foliar diseases varies; it may be on the order of several months after infection. Lower temperatures increase the latency of the disease. Citrus canker bacteria can stay viable in old lesions and other plant surfaces for several months.

Citrus canker lesions on fruit.

Pathogenicity

Xanthomonas axonopodis has the capability to form a biofilm for attachment on the host. The biofilm is the result of the production of extracellular polysaccharides (xanthan). The biofilm ensures the virulence and epiphytic survival of *X. axonopodis* pv. *citri* prior to the development of citrus canker. In addition, the bacteria secrete transcriptional activator-like (TAL) effectors through type III secretion system. The effector interacts with host machinery to induce transcription for genes that regulate plant hormones such as gibberellin and auxin.

Disease Cycle

Xanthomonas axonopodis pv. *citri* overseason in infected area which is canker lesion on leaf or stem. The bacteria ooze out of the lesions when there is free moisture. During the rainy weather, wind-blown rain carries the inoculum to the new susceptible hosts. The bacteria infect new plants through stomata and wounds. The wound can be caused by pruning or hedging that could cut open mesophyll tissues for direct infection. The rain can also cause water congestion on leaf surface, form column of water through stomata and promote infection through the natural opening. The infection can form on fruit, foliage and young stem. Leaves and stems are most susceptible to infection within the first six weeks of initial growth. Infection of fruit is most likely to occur during the 90 day period after petal fall during fruit formation. The varied size of lesions on citrus fruit is because of the multiple cycle of infections and can reflect different-aged lesions on the same fruit.

Favorable Environmental Condition

Wind-driven rain plays major role in the dispersal of *X. axonopodis*. The bacteria are said to be readily dispersed by splashed rain and wind and the quantity of *X. axonopodis* declines after the first event of wind-blown rain dispersal. Apart from that, the bacteria also favor warm weather. The cases of citrus canker are more acute in areas that receive high rainfall and high mean temperature such as Florida. Often, cankers emerge briskly during fall, slowly during winter and most rapidly in mid to late spring.

Detection

The disease can be detected in groves and on fruit by the appearance of lesions. Early detection is critical in quarantine situations. Bacteria can be tested for pathogenicity by inoculating multiple citrus species with them. Additional diagnostic tests (antibody detection), fatty-acid profiling, and genetic procedures using polymerase chain reaction can be conducted to confirm diagnosis and may help to identify the particular canker strain. Clara H. Hasse determined that citrus canker was not of fungoid origin but caused by bacterial parasite. Her research published in the 1915 played a major part in saving citrus crops in multiple states.

Susceptibility

Not all species and varieties of citrus have been tested for citrus canker. Most of the common species and varieties of citrus are susceptible to it. Some species are more susceptible than others, while a few species are resistant to infection.

Susceptibility	Variety
Highly susceptible	Grapefruit (Citrus x paradisi), Key lime (C. aurantiifolia), Pointed leaf hystrix (C. hystrix), lemon (C. limon)
Susceptible	Limes (C. latifolia) including Tahiti lime, Palestine sweet lime; trifoliate orange (Poncirus trifoliata); citranges/citrumelos (P. trifoliata hybrids); tangerines, tangors, tangelos (C. reticulata hybrids); sweet oranges (C. sinensis); bitter oranges (C. aurantium)
Resistant	Citron (C. medica), Mandarins (C. reticulata)
Highly resistant	Calamondin (X Citrofortunella), kumquat (Fortunella spp.)

Management

Quarantine measures are implemented in areas where citrus canker is not endemic or has been obliterated to prevent the introduction of *X. axonopodis*. On the other hand, in the regions where citrus canker outbreak occurs, Integrated Pest Management (IPM) is utilized. The significant features of this management program is the transposition of susceptible citrus plants to field resistant citrus cultivars. Apart from using the resistant cultivars in fields, there are several measures that are taken to control citrus canker from causing failed crop. The measures can be divided into three major categories which are exclusion, eradication and sanitation.

Exclusion

In exclusion method, citrus trees or fruits from outside of the country are inspected to ensure they are bacterial-free trees. Under the management program, the production of *Xac* (*X. axonopodis* pv. *citri*)-free nursery trees for exclusion of canker from orchard is also mandatory. Because the bacteria can be introduced from the countries with canker issue, strict restrictions on the citrus importation are implemented in citrus-growing countries. Citrus trees will only be grown on canker-free fields for at least one year after effective eradication. The planting sites are also chosen to minimize the favorable environmental condition for the introduction of *X. axonopodis*. For example, areas with strong wind are avoided to evade the dispersal of bacterial inoculum to the susceptible citrus trees.

Eradication

Once citrus canker is introduced into a field, removal of the infected trees is enacted to halt further spread of the bacteria. For instance, in Florida, all citrus trees within 579 m of infected trees must be eradicated. In the process, the infected trees are uprooted and burned. In urban areas, the trees are cut down and chipped, then disposed in the landfills.

Sanitation

X. axonopodis pv. *citri* can be transmitted by mechanical means such as humans and machinery. As a sanitation measure, the workers in citrus orchards are required to do thorough decontamination of personnel and equipment to prevent the spread of bacteria from the infected areas. Aerosol inoculum is able to cause infection in wetted foliage in the zone of bacterial dispersal. Vehicles can also become contaminated by contacting the wet foliage. The contaminated equipments and machines can be disinfected by spraying bactericidal compound.

VIRAL DISEASES

Tobacco Mosaic Virus

Tobacco mosaic virus (TMV) is a positive-sense single stranded RNA virus in genus *Tobamovirus* that infects a wide range of plants, especially tobacco and other members of the family Solanaceae. The infection causes characteristic patterns, such as mosaic-like mottling and discoloration on the leaves (hence the name). TMV was the first virus ever to be discovered. Although it was known from the late 19th century that a non-bacterial infectious disease was damaging tobacco crops, it was not until 1930 that the infectious agent was determined to be a virus. It is the first pathogen identified as a virus.

Structure

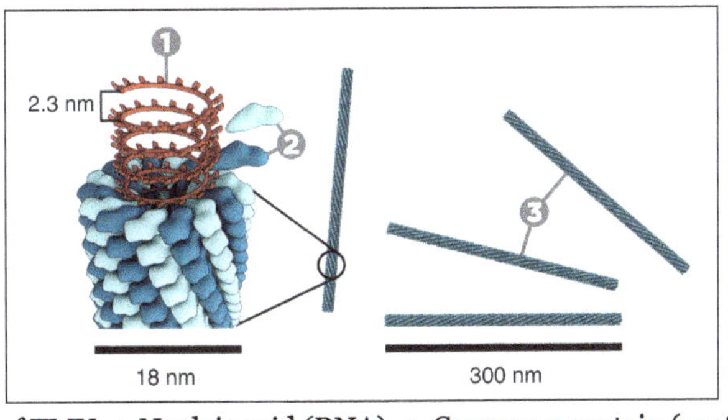

Schematic model of TMV: 1. Nucleic acid (RNA), 2. Capsomer protein (protomer), 3. Capsid.

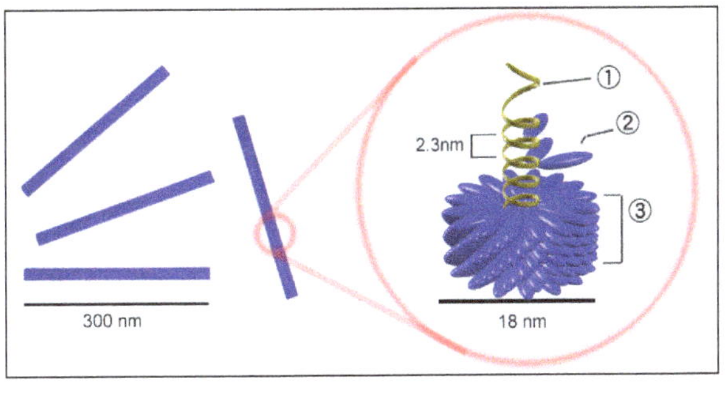

Tobacco mosaic virus has a rod-like appearance. Its capsid is made from 2130 molecules of coat protein and one molecule of genomic single strand RNA, 6400 bases long. The coat protein self-assembles into the rod-like helical structure (16.3 proteins per helix turn) around the RNA, which forms a hairpin loop structure. The protein monomer consists of 158 amino acids which are assembled into four main alpha-helices, which are joined by a prominent loop proximal to the axis of the virion. Virions are ~300 nm in length and ~18 nm in diameter. Negatively stained electron microphotographs show a distinct inner channel of ~4 nm. The RNA is located at a radius of ~6 nm and is protected from the action of cellular enzymes by the coat protein. There are three RNA nucleotides per protein monomer. X-ray fiber diffraction structure of the intact virus was studied based on an electron density map at 3.6 Å resolution. Inside the core helix, coiled RNA molecule is present, which is made up of nearly 6500 nucleotides.

Genome

The TMV genome consists of a 6.3–6.5 kb single-stranded (ss) RNA. The 3'-terminus has a tRNA-like structure, and the 5' terminus has a methylated nucleotide cap. (m7G5'pppG). The genome encodes 4 open reading frames (ORFs), two of which produce a single protein due to ribosomal readthrough of a leaky UAG stop codon. The 4 genes encode a replicase (with methyltransferase [MT] and RNA helicase [Hel] domains), an RNA-dependent RNA polymerase, a so-called movement protein (MP) and a capsid protein (CP).

Physicochemical Properties

TMV is a thermostable virus. On a dried leaf, it can withstand up to 50 °C (120 degree Fahrenheit) for 30 minutes.

TMV has an index of refraction of about 1.57.

Disease Cycle

TMV does not have a distinct overwintering structure. Rather, it will over-winter in infected tobacco stalks and leaves in the soil, on the surface of contaminated seed (TMV can even survive in contaminated tobacco products for many years). With the direct contact with host plants through its vectors (normally insects such as aphids and leafhoppers), TMV will go through the infection process and then the replication process.

Infection and Transmission

After its multiplication, it enters the neighboring cells through plasmodesmata. The infection spreads by direct contact to the neighboring cells, for its smooth entry, TMV produces a 30 kDa movement protein called P30 which enlarges the plasmodesmata. TMV most likely moves from cell-to-cell as a complex of the RNA, P30, and replicate proteins.

It can also spread through phloem for longer distance movement within the plant. Moreover, TMV can be transmitted from one plant to another by direct contact. Although TMV does not have defined transmission vectors, the virus can be easily transmitted from the infected hosts to the healthy plants, by human handling.

Replication

Following entry into its host via mechanical inoculation, TMV uncoats itself to release its viral [+] RNA strand. As uncoating occurs, the MetHel:Pol gene is translated to make the capping enzyme MetHel and the RNA Polymerase. Then the viral genome will further replicate to produce multiple mRNAs via a [-]RNA intermediate primed by the $tRNA_{HIS}$ at the [+]RNA 3' end. The resulting mRNAs encode several proteins, including the coat protein and an RNA-dependent RNA polymerase (RdRp), as well as the movement protein. Thus TMV can replicate its own genome.

After the coat protein and RNA genome of TMV have been synthesized, they spontaneously assemble into complete TMV virions in a highly organized process. The protomers come together to form disks or 'lockwashers' composed of two layers of protomers arranged in a helix. The helical capsid grows by the addition of protomers to the end of the rod. As the rod lengthens, the RNA passes through a channel in its center and forms a loop at the growing end. In this way the RNA can easily fit as a spiral into the interior of the helical capsid.

Host and Symptoms

Tobacco mosaic virus symptoms on tobacco.

Tobacco mosaic virus symptoms on orchid.

Like other plant pathogenic viruses, TMV has a very wide host range and has different effects depending on the host being infected. *Tobacco mosaic virus* has been known to cause a production loss for flue cured tobacco of up to two percent in North Carolina. It is known to infect members of nine plant families, and at least 125 individual species, including tobacco, tomato, pepper (all members of the useful Solanaceae), cucumbers, and a number of ornamental flowers. There are many different strains. The first symptom of this virus disease is a light green coloration between the veins of young leaves. This is followed quickly by the development of a mosaic or mottled pattern of light and dark green areas in the leaves. Rugosity may also be seen where the infected plant leaves display small localized random wrinkles. These symptoms develop quickly and are more pronounced on younger leaves. Its infection does not result in plant death, but if infection occurs early in the season, plants are stunted. Lower leaves are subjected to mosaic burn especially during periods of hot and dry weather. In these cases, large dead areas develop in the leaves. This constitutes one of the most destructive phases of *Tobacco mosaic virus* infection. Infected leaves may be crinkled, puckered, or elongated. However, if TMV infects crops like grape and apple, it is almost symptomless.

Environment

TMV is known as one of the most stable viruses. It has a very wide survival range. As long as the surrounding temperature remains below approximately 40 °C, TMV can sustain its stable form.

All it needs is a host to infect. If necessary, greenhouses and botanical gardens would provide the most favorable condition for TMV to spread out, due to the high population density of possible hosts and the constant temperature throughout the year.

Treatment and Management

One of the common control methods for TMV is sanitation, which includes removing infected plants and washing hands in between each planting. Crop rotation should also be employed to avoid infected soil/seed beds for at least two years. As for any plant disease, looking for resistant strains against TMV may also be advised. Furthermore, the cross protection method can be administered, where the stronger strain of TMV infection is inhibited by infecting the host plant with mild strain of TMV, similar to the effect of a vaccine.

In the past ten years, the application of genetic engineering on a host plant genome has been developed to allow the host plant to produce the TMV coat protein within their cells. It was hypothesized that the TMV genome will be re-coated rapidly upon entering the host cell, thus it prevents the initiation of TMV replication. Later it was found that the mechanism that protects the host from viral genome insertion is through gene silencing.

Scientific and Environmental Impact

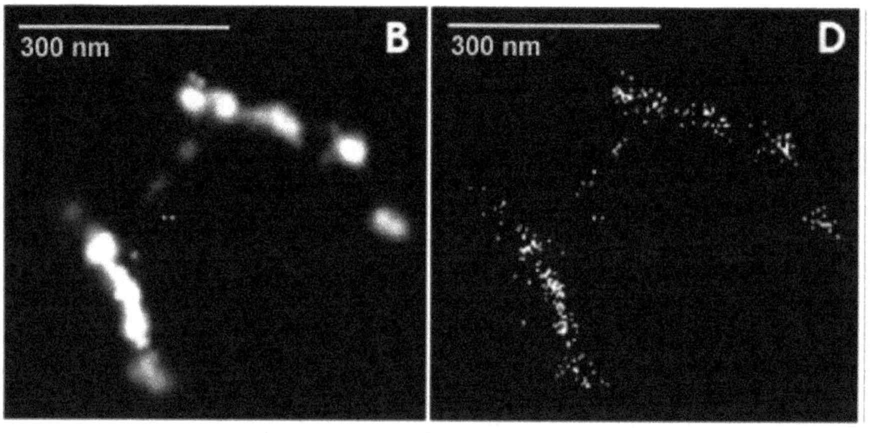

TMV virus: Super resolution light microscopy.

The large amount of literature about TMV and its choice for many pioneering investigations in structural biology (including X-ray diffraction), virus assembly and disassembly, and so on, are fundamentally due to the large quantities that can be obtained, plus the fact that it does not infect animals. After growing a few infected tobacco plants in a greenhouse and a few simple laboratory procedures, a scientist can easily produce several grams of the virus.

Investigational Uses

Due to its cylindrical shape, high aspect-ratio, self-assembling nature, and ability to incorporate metal coatings (nickel and cobalt) into its shell, TMV is an ideal candidate to be incorporated into battery electrodes. Addition of TMV to a battery electrode increases the reactive surface area by an order of magnitude, resulting in an increase in the battery's capacity by up to six times compared to a planar electrode geometry.

Tomato Spotted Wilt Virus

Tomato spotted wilt virus (*TSWV*) is a spherical negative-sense RNA virus that has a diameter between 80-110nm.

Transmission and Lifespan

TSWV, which is transmitted by thrips, causes serious losses in economically important crops and it is one of the most economically devastating plant viruses in the world. The circulative propagative transmission of *TSWV* is carried out by at least ten different species of thrips. The most common species is *Frankliniella occidentalis* (western flower thrips) as it is the vector that predominantly transmits *TSWV* globally and in greenhouses. The rapid developmental and reproductive rate of the thrips contributes to the spread of *TSWV*. The amount of time it takes for insects to acquire the virus (acquisition period) and the amount of time it takes for the virus to move from the insect to the plant (inoculation) for *TSWV* varies depends on the vector species. For *Frankliniella occidentalis*, the acquisition and inoculation of *TSWV* can be as short as 5 minutes. However, the acquisition and inoculation periods for optimal transmission is 21.3 hours and 42.7 hours, respectively.

Transmission of *TSWV* can only occur when larvae stage thrips acquire *TSWV*. The larvae stage for thrips lasts around 1-3 days. *TSWV* is acquired by thrips when they feed on infected plants. Adult thrips cannot be infected with *TSWV* as their midgut barrier successfully prevents infection. However, thrips that have successfully become infected with *TSWV* in the larvae stage can transmit the virus throughout their lifetime. In order to protect their eggs, thrips insert their eggs into various types of plant tissue. Eggs can be found in the stems, leaves, or flowers of plants. Thrips hatch in 2-3 days and complete their lifecycle in 20-30 days. Adult thrips feed on the flower bud, stem and leaf parts of the plant.

Hosts and Symptoms

TSWV infects a variety of hosts, contributing to its global economic impact on crops. There are over a thousand different hosts for *TSWV*. The host range of *TSWV* includes agronomically important crops like tomatoes and tobacco. The symptoms of *TSWV* vary from host to host. There is also variability of symptoms within a single type of host due to the age of the plant, nutrition and the environment (especially temperature). Common symptoms include stunting, ringspots on fruit and necrosis of leaves. There are many different strains of *TSWV*, and differences in symptoms may also be attributed to the differences in the number of strains present.

Management

Prevention is key in managing *TSWV*. Once a plant becomes infected with *TSWV*, there are no practical ways to cure the virus infected plant. The most efficient method of containing this disease is genetic resistance. There are several different resistance genes identified in multiple crops. In some crops the resistance genes have been effective, however, in others, some strains of *TSWV* have been discovered to overcome the resistance gene, such as the *Sw-5* resistance gene in tomato. The *Sw-5* resistance gene in tomato is a dominant resistant gene. The *Sw-5* gene gives resistance to the *TSWV* through a hypersensitive response. A hypersensitive response is when the plant cells that surround the infection undergo cell death which would then deprive the virus of the cell machinery it needs to replicate and infect the plant further. There have been several strains of *TSWV* detected in countries such as Australia, Spain and the United States that can overcome the *Sw-5* resistant gene. However, those strains of *TSWV* have not been spread worldwide so the *Sw-5* gene is still useful.

Other important prevention techniques include buying virus and thrips-free transplants and managing thrips populations. Introducing species that naturally prey on thrips, such as the minute pirate bugs (*Orius insidious*) and big eyed bugs (*Geocoris punctipes*), may help reduce transmission of *TSWV*. Insecticides are not an efficient way to decrease the vector population because the vectors rapidly develop resistance. Removing weeds and infected plants is a good way to prevent more infections in the greenhouse. Sanitation practices such as the destruction or removal of old crops by plowing or physical removal are often used in the field.

Cucumber Mosaic Virus

Cucumber mosaic virus (CMV) is a plant pathogenic virus in the family *Bromoviridae*. It is the type member of the plant virus genus, *Cucumovirus*. This virus has a worldwide distribution and a very wide host range. In fact it has the reputation of having the widest host range of any known plant virus. It can be transmitted from plant to plant both mechanically by sap and by aphids in a stylet-borne fashion. It can also be transmitted in seeds and by the parasitic weeds, *Cuscuta sp.* (dodder).

Hosts and Symptoms

In plant tissue this virus makes characteristic viral inclusion bodies which can be diagnostic. They are hexagonal in shape and stain both in a protein stain and a nucleic acid stain. The inclusions can also be rhomboidal, may appear hollow and can form larger aggregates. The inclusions are not uniformly distributed and can be found in epidermal, mesophyll, and stomatal cells. These inclusions are made up of virus particles.

This virus was first found in cucumbers (*Cucumis sativus*) showing mosaic symptoms in 1934, hence the name *Cucumber mosaic*. Since it was first recognized, it has been found to infect a great variety of other plants. These include other vegetables such as squash, melons, peppers, beans, tomatoes, carrots, celery, lettuce, spinach and beets, various weeds and many ornamentals and bedding plants, such as *Narcissus*. Symptoms seen with this virus include leaf mosaic or mottling, yellowing, ringspots, stunting, and leaf, flower and fruit distortion.

CMV shows symptoms on leaves known as the “shoestring” effect for most host species. This effect causes young leaves to appear narrow and the entire plant to be stunted.

Specifically CMV can cause cucumbers to turn pale and bumpy. The leaves of these plants turn mosaic and their rugosity is often changed, making leaves wrinkled and misshapen. Growth of these plants is usually stunted and produces few flowers. Often cucumber fruits are oddly shaped and appear gray. This appearance often leads to cucumbers being referred to as “white pickles”. Often infected cucumbers are bitter.

In celery, CMV can cause streaking and spotting and can be often confused with symptoms of the celery mosaic virus.

Symptoms of CMV in lettuce can be similar to those of lettuce mosaic virus. Infected plants show symptoms of chlorosis, stunting and often do not properly head.

Some of the most important fruits and vegetables affected by CMV are peppers, bananas, tomatoes and cucurbits.

CMV in peppers causes slightly different symptoms than the previously mentioned. Pepper plants often have severe foliar damage, shown as mosaic and necrotic rings. Often the peppers themselves are misshapen and contain chlorotic rings and spots.

Tomato plants are usually stunted and have poorly shaped leaves, or "fernleaf", when infected by CMV. Also certain strains of CMV can cause partial or total crop loss.

The cucumber mosaic virus has been found on American beautyberry, an important wildlife and pollinator food source plant native to North America.

Symptoms in *Commelina diffusa*.

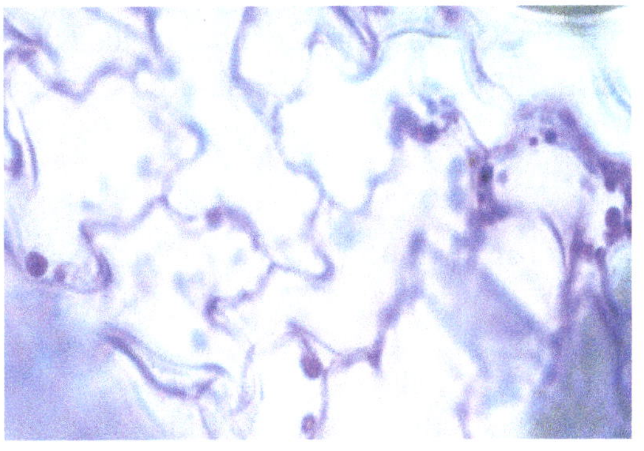

Cucumber mosaic virus inclusion bodies.

Disease Cycle

CMV is mainly transmitted by aphids, but it can also be mechanically spread by humans in some cases. However, the mechanically spreading of this virus is not as common as the case of other virus (such as Tobacco Mosaic Virus, TMV), because CMV is not a very stable virus. When it is transmitted by aphids, this virus has an acquisition period of five to ten seconds and an inoculation period of about a minute. Nevertheless, after two minutes, the probability of inoculation largely decreases, and within two hours it is practically impossible to transmit it. Moreover, CMV can overwinter in perennial plants and weeds, as it can survive the winter in the roots of the plant and move to the aerial parts in spring, where it can be transmitted by aphids to other plants.

Once the virus penetrates into the host cell, it releases its RNAs into the host cytoplasm. Then, proteins 1a and 2a are produce to enable the virus replication, which takes place in viral factories, which are subcellular compartments which increase the efficiency of this process. There, a dsRNA genome is synthetized from the ssRNA(+) and transcribed in order to obtain viral mRNAs as well as new ssRNAs. Afterwards, the capsid proteins are produced and the new viral particles are assembled. Finally, the virus is ready to move to a new cell by triggering the formation of tubular structures which mediate the movement of the virions. The short-distance (cell-to-cell) movement of the virus is achieved via plasmodesmata, while the long-distance one (within the plant) occurs via the phloem.

Properties

Genome

CMV is a linear positive-sense, tripartite single-stranded RNA virus. Its genome size is 8.623 kb and it is divided among RNA1 (3357 pb), RNA2 (3050 pb) and RNA3 (2216 pb), all of which has a tRNA-like structure. These three RNAs encode five proteins, proteins 1a, 2a, 2b, movement protein (MP) and coat protein (CP). While proteins 1a and 2a are responsible for the replication of the virus, protein 2b is the host-silencing suppressor.

Its total genome size 8.621 kb and is broken into three parts. The largest part is 3.389 kb; the second largest is 3.035 kb; the third largest is 2.197 kb. The RNA is surrounded by a protein coat consisting of 32 copies of a single structural protein which form isometric particles.

Virion

This virus presents non-enveloped, icosahedral or bacilliform virions of 26-35 nm in diameter. The different RNAs are encapsidated in distinct particles, which results in a variety of virions.

Environment

CMV is naturally found in temperate areas, where aphids, one of its main vectors, are also found.

Diagnosis

The presence of this virus in a plant can be confirm by serological (ELISA), genetic (PCR) or host range tests.

Management

Nowadays, there is not any chemical capable of removing this virus from an infected plant. Therefore, the best control in this case is prevention of the infection and eradication. To achieve this, it is crucial to remove weeds and diseased plants from the field, as well as use clean and sanitized tools. Another option consists of the use of resistant varieties or the so-called “trap crops”.

Potato Virus Y

Potato virus Y (PVY) is a plant pathogenic virus of the family *Potyviridae*, and one of the most important plant viruses affecting potato production.

PVY infection of potato plants results in a variety of symptoms depending on the viral strain. The mildest of these symptoms is production loss, but the most detrimental is ‘potato tuber necrotic ringspot disease’ (PTNRD). Necrotic ringspots render potatoes unmarketable and can therefore result in a significant loss of income. PVY is transmissible by aphid vectors but may also remain dormant in seed potatoes. This means that using the same line of potato for production of seed potatoes for several consecutive generations will lead to a progressive increase in viral load and subsequent loss of crop.

An increase in potato plant infection with viruses over the past few years has led to considerable losses to the South African potato industry. The increased rate of infection may be attributed to several factors. These include a marked decrease in the effectiveness and administration of chemicals used in vector control, the use of infected seed potatoes in cultivation, incorrect irrigation and farming methods as well as a lack of a sensitive, rapid and reliable method of detection. An increase in the average temperature of winters as a consequence of global warming has also led to an increase in aphid numbers, which in turn has led to an increase in viral distribution.

Hosts, Strains and Symptoms

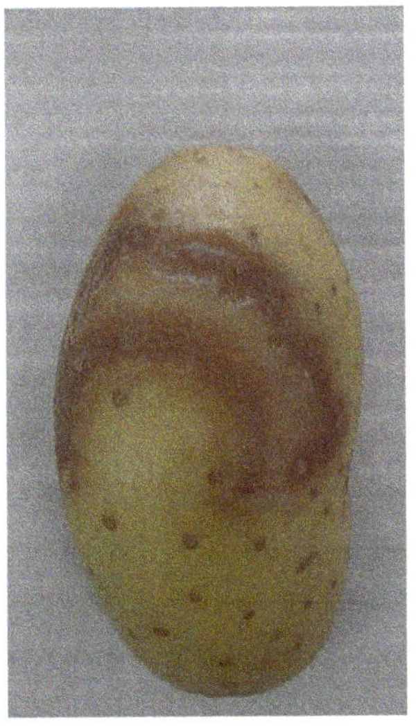

Potato illustrating necrotic ringspot disease.

PVY belongs to the genus *Potyvirus*, of which it is the type member. *Potyvirus* is the largest genus of plant viruses, and possibly the most destructive one in potato crops. The genus includes more than 200 species that bring about significant losses in the agricultural arena. PVY infects many economically important plant species. These include potato (*Solanum tuberosum*), tobacco (*Nicotiana tabacum*), tomato (*Solanum lycopersicum*) and pepper (*Capsicum* spp.). The level of damage to crop is determined by the strain of PVY infecting the plants, the viral load, the time at which infection occurs as well as the tolerance the host possesses toward the virus. Resistance to PVY infection by hosts is low in many cases. Infection of a potato field with PVY may ultimately result in 10-100% loss in yield.

It has been shown that the PVY has different isolates according to the symptoms they induce in various potato plant species. Extensive biological, serological and molecular variability of PVY isolates makes the classification of isolates as particular strains particularly difficult. Occurrence of a variety of symptoms and the emergence of the necrotic PVY^{NTN} has led to a search for more reliable classification tools than simple serological identification. Traditionally three chief strains of PVY are recognized: PVY^{C}, PVY^{N} and PVY^{O}. PVY^{C}, originally known as *Potato Virus C*, was the first to be recognized and was identified in the 1930s. PVY^{C} induces hypersensitive responses in a wide range of potato cultivars. These reactions include the formation of mild mosaic patterns or stipple streak. Unlike the other strains of PVY, some PVY^{C} strains are non-aphid transmissible. Previous studies by Visser *et al.* did not identify any of the local isolates as being PVY^{C} but it has been reported to occur to in South Africa. A second strain of PVY is PVY^{N}. Some notes on suspected variant of Solanum virus 2 (*Potato virus Y*). This strain was described in tobacco plants growing close to potato plants. PVY^{N} results in leaf necrosis and mild or even no damage to the tubers. The ordinary strain of PVY is denoted as PVY^{O}. Infection of a potato plant with the PVY^{O} strain results in mild tuber damage and does not cause leaf necrosis. Both PVY^{N} and PVY^{O} are aphid transmissible and occur in South Africa. In Europe these two strains have been shown to have recombined to form PVY^{NTN}. The PVY^{NTN} has been accredited with the ability to induce potato tuber necrotic ringspot disease (PTNRD). Tubers damaged by PTNRD become unmarketable and infection by PVY^{NTN} thus results in a larger economic impact than infection by the other strains.

Transmission

PVY may be transmitted to potato plants through grafting, plant sap inoculation and through aphid transmission. The most common manner of PVY infection of plant material in the field is through the aphid and although aphids on their own can directly damage potato plants it is their role as viral vectors which has the greatest economic impact. In cold climates aphids spend the winter either as wingless aphids giving birth to live young (viviparae) or as eggs. Hosts such as weeds and other crops serve as breeding grounds for these aphids and form a temporary area of colonization before the aphids migrate to the potato fields. In moderate climates, such as in South Africa, aphids are thought to reproduce asexually on weeds, other crops, indigenous plants and garden plants. This means that there are a number of aphids present year-round. The importance in effective and stringent monitoring of aphid populations is stressed in a review by Radcliffe and Ragsdale as PVY virions are introduced to potato fields almost solely by winged aphids from a virus source outside these fields. Wingless aphids have not yet been linked to the spread of PVY in potato fields.

The green peach aphid (*Myzus persicae*) has been found to be most effective in its role as viral vector, but others such as *Aphis fabae*, *Aphis gossypii*, *Aphis nasturtii*, *Macrosiphum euphorbiae*, *Myzus (Nectarosiphon) certus*, *Myzus (Phorodon) humuli* and *Rhopalosiphum insertum* are also strongly associated with viral transmission. The Agricultural Research Council-Vegetable and Ornamental Plant Institute (ARC-VOPI) 6 of South Africa identified twenty five species of aphid able to function as PVY vectors. The efficiencies of some of these aphids to function as PVY vectors were also established and were found to vary between the different species. In South Africa, *Aphis fabae*, *Aphis gossypii* and *Aphis nasturtii* are the most common and efficient PVY vectors found in the field. Apart from being classed according to efficiency as vectors, aphids can also be divided into two subgroups, namely colonizing and non-colonizing species. Colonizing aphids are aphids which reproduce and establish themselves on potato plants, specifically, while non-colonizing aphids do not reproduce nor establish colonies on potato plants. Colonizing aphids are better adapted to life on potato plants and are thus generally considered as better PVY vectors than non-colonizing aphids. Noncolonizing aphids do not primarily feed on potato plants but do occasionally feed on them while searching for a more suitable host. Their lower efficiency as PVY vector is cancelled out by the sheer numbers in which they occur. Because of this, all aphids present in and around potato fields must be considered as possible vectors and their numbers carefully monitored.

Transmission of PVY by aphids occurs in a non-persistent, non-circulative manner which suggests a less intimate interaction between virion and vector than is the case of circulative virions. The fact that the virions are transmitted in a non-persistent fashion means that viral replication does not occur within the aphid vector and that, unless the aphid feeds on infected plants, it loses its ability to infect plants after two to three feedings. The virions attach to the aphid stylet in a matter of seconds and may remain infectious for four to seventeen hours. The distance over which the virions can be transmitted is limited due to the short period for which they remain infectious. Although the short life span outside plants inhibits long-distance viral transmission, it does not reduce the transmission efficiency bestowed by the quick rate of viral acquisition and inoculation within a field.

Upon entrance into the plant cell, the virus coat protein disassembles and releases its RNA genome. The viral RNA serves as mRNA, and although little is known about the translation thereof, it is believed that the 5' non-coding region functions as an enhancer of translation. The translated mRNA results in a polyprotein which is processed into mature proteins. Each polyprotein is then cleaved into ten different proteins which are believed to be multifunctional. These proteins, along with host proteins, assemble to form a replication complex. This complex performs negative-strand RNA synthesis, using the positive strand of viral RNA as a template. Once the additional RNA copies have been produced, they code for the synthesis of various proteins, as mentioned before, as well as coat proteins. These coat proteins will now enclose the newly formed genomes to give rise to new virions. It has been suggested that enclosure of the newly formed virions is initiated by the interaction of the coat proteins with the 5'terminus and that the coat protein is built up towards the 3'terminus. The entire process of viral replication occurs within the endoplasmic reticulum. These newly synthesized viral particles are subsequently transported through the plasmodesmata to adjacent plant cells via several assisting potyvirus proteins. Distribution of viruses within the plant occurs according to the source-sink relationship between maturing and growing tissues. Virus concentration throughout the plant is high and this greatly increases the chance of uptake by aphids. Infection of plants by potyviruses can be varied in the symptoms shown. Infection can include veinal necrosis, mosaic symptoms as well as leaf malformation. Infected plants that do not show symptoms may have infected canopies and will yield lower quality products than their healthy counterparts.

Potato – PVY^{NTN} Interaction

Since PVY^{NTN} causes great loss in potato production, the research of potato – potato virus Y^{NTN} interaction is important. Sensitive potato cultivars respond to PVY^{NTN} inoculation with development of typical symptoms. On inoculated leaves 5 – 7 days after inoculation chlorotic and necrotic ringspots develop. As the virus spreads through the plant the systemic symptoms develop on uninoculated leaves. 10 days after inoculation wrinkles and mosaic chlorosis appear, leading to a palm tree appearance (leaf drop).

The viral defense mechanisms of plants will primarily try to restrict the movement of the virus. In failing this, it may attempt to induce cell death in infected tissue, thereby preventing the spread of virions. Although the precise mechanism of disease induction by potyviruses in plants is unknown, it is known that these viruses cause a significant shutdown of host gene expression during viral replication.

Physiological changes in potato plants as a response to PVY^{NTN} infection were intensively studied. At early stages of infection, meaning first 12 hours, photosynthesis related genes, genes involved in perception, signalling and defence response were shown to be differentially expressed. 24 h after inoculation the amount of salicylic acid increased.

A disruption in gene expression disrupts the normal cellular function of cells which could be the cause of the physical symptoms that the plant demonstrates. At the time of symptoms development, research on interaction between susceptible potato cultivar and PVY^{NTN} showed changes in cytokinin level. In inoculated leaves showing symptoms modifications in chloroplast structure and size, lower chlorophyll levels and differential activity of soluble and ionically-bound peroxidases were detected.

At later stages of PVY^{NTN} infection total protein concentration increased in sensitive potato cultivar while no such pronounced changes were observed in tolerant and moderately tolerant potato cultivars. Gene expression studies revealed changes in expression of genes for heat-shock proteins, catalase, β-1, 3-glucanase and genes involved in photosynthesis.

Molecular Description of Potato Virus Y

Potyvirus virions consist of non-enveloped filamentous structures that are 680 – 900 nm in length and 11 to 15 nm in width. Morphologically the potyvirus capsid consists of approximately 2000 copies of coat protein (CP).

The capsid encapsulates a single strand of positive sense RNA which is in the order of 10 kb in length and has a non-translated 5'-terminal region (5'-NTR) as well as a 3'-poly-A tail. The positive sense genome contains a single extended open reading frame and acts directly as mRNA. The 144 nucleotide 5'-NTR is particularly rich in adenine residues and has very few guanine residues. Rather than a conventional cap structure, the 5'NTR is associated with a Viral genome linked protein (VPg) which is said to act as an enhancer of transcription.

The 5'-leader sequence has an internal ribosome entry site (IRES) and cap-independent translation regulatory elements (CIREs). The IRES directs cap-independent translation through a mechanism similar to that used by eukaryotes. The extended open reading frame encodes for a 350 kDa polyprotein. This polyprotein is proteolytically processed by viral proteases (NIa, HC-Pro and P1) and undergoes co- and post-translational cleavage to yield several multi-functional proteins. These include the following: P1 (P1 Protein), HCPro (Helper Component Proteinase), P3 (P3 Protein), 6K1 (6-kDa Protein 1), CI (Cylindrical Inclusion), 6K2 (6-kDa Protein 2), VPg (Viral Protein genome-linked), NIaPro (Nuclear Inclusion Protein a, Proteinase domain), NIb (Nuclear Inclusion Protein b) and the CP (Coat Protein).

Diagnostic Techniques for Detection of Potato Virus Y

ELISA

In the past, crops were inspected visually to determine whether or not they were disease free. Visual inspection was also used as a basis for seed certification. Determination of viral status through visual inspection is incredibly difficult as the symptoms may be masked or the infection latent. As a result, post season tests and inspections were introduced. These tests involved the cultivation of previously harvested material in greenhouses. The resulting plants were inspected for a more accurate estimate of viral status. Although this method of screening did offer some degree of monitoring of viral presence it was subjective and highly ineffective. Enzyme-linked immunosorbent assay (ELISA) screening of crops and seed potatoes replaced visual inspection in the early 1970s. The use of ELISA offered routine diagnostic laboratories a quick, effective and sensitive method of screening for a wide range of potato plant viruses.

Detection of pathogens using ELISA relies on the interaction between the antigen and specific antibodies and has become a popular and cost-effective means of routine detection. In an ELISA the solid phase can be coated with the sample of interest containing the antigen. The efficiency to which the antigen binds to the solid phase is dependent on temperature, length of exposure as well as concentration. Solid phases used include nitrocellulose membranes, paper, glass, agarose and polystyrene or polyvinylchloride microtiter plates. Microtiter plates are the most widely used solid phase because they are easy to handle, allow for automation and for analysis using microtiter plate readers. A drawback of these plates is that they are highly absorptive and this increases the incidence of non-specific binding of components used in the ELISA. Non-specific binding to the plates is reduced through the use of buffers containing proteins such as casein and non-ionic detergents such as Tween 20. After coating, excess sample is removed and the plate typically treated with a 1% casein containing solution. Subsequent to this the solid phase is treated with antibodies raised against the antigen of interest. After each incubation step the plate is washed with Tween 20 containing PBS. These washing steps are aimed to wash away any non-specifically bound components.

Nonspecifically bound components are less strongly bound than the specific bound ones. Detection is achieved either through the addition of an enzyme-coupled antibody or the addition and detection of a biotinylated antibody. In a system using an enzyme-coupled antibody the subsequent addition of an appropriate substrate results in the formation of a colour proportional to the amount of antigen. Alternatively the plate can be coated with antibody followed by incubation with the sample that is to be detected. This, in turn, can be detected as described above and is then referred to as the double antibody sandwich (DAS) ELISA. Both of these systems, however, have a disadvantage in that coupling of the enzyme to the antibody may result in steric hindrance which in turn may result in a loss in function of the antibody and/or the enzyme. This may be overcome through the use of a biotin-avidin or biotin-streptavidin bridge. In this type of system biotin is coupled to the antibody. The biotin molecule has no influence on the working of the antibodies and is easily detectedusing avidin or streptavidin conjugated to a suitable enzyme. Streptavidin has an extremely high affinity for biotin which results in even a higher degree of specificity than a system in which the enzyme is coupled directly the antigen. To establish whether or not the antigen is present, a substrate specific for the enzyme used is added. The enzyme then converts the substrate to a coloured product and the colour intensity can be correlated to the amount of antibodies bound and thus the amount of antigen present. A DAS-ELISA has the advantage that it can increase the specificity of the ELISA and reduce the occurrence of non-specific binding. As a result, the DAS-ELISA principle is commonly employed in ELISA's for the detection of plant pathogens in plant sap without prior purification of the pathogen.

The ELISA is considered to be a safe, inexpensive and rapid method for detection of plant viruses. The inexpensive nature and relative simplicity thereof allows for it to be used as a workhorse within the agricultural sector and is used to screen thousands of samples per year. Unfortunately ELISAs are not completely failsafe. Virus levels within potato tubers, which are screened by ELISA for use as seed potatoes, are normally low while the tubers are dormant. ELISA detection of viruses in these potatoes is difficult and absorbance values may fall below the set cut-off value. For this reason, seed tuber screening is performed on sprouting rather than dormant tubers. Although this results in more reliable readings than direct tuber testing, it does delay the certification of seed potatoes. Another disadvantage of an immuno-based detection method is that changes at the gene level may have an influence on the immunogenicity of the antigen to be detected. In terms of potato plant viruses, mutations within the CP gene may cause the CP to undergo conformational changes rendering antibodies produced against the previously present virus less effective.

RT-PCR

Reverse transcriptase polymerase chain reaction (RT-PCR) has become a powerful and effective method for detection of potato plant viruses within potato plant material and even dormant potatoes. Only a minute piece of plant material is required for analysis using RT-PCR. Considering the protocol described within this thesis, 0.1 g of plant material is enough for 14,500 separate reactions. During a RT-PCR specific target RNA sequences are amplified exponentially into DNA copies. For this to occur, however, the RNA of the virus must first be transcribed to DNA by means of a reverse transcriptase polymerase. This polymerase synthesizes a DNA strand using the RNA as template. This results in a DNA/RNA complex. For synthesis of a DNA strand from the RNA template only the reverse primer is required since the RNA is a single strand arranged from 5' to 3'. Subsequently, the newly synthesized DNA strand is used as a template for traditional PCR.

Different types of reverse transcriptase polymerases are available to suit different needs and reaction conditions. Reverse transcriptase enzymes commonly used include AMV RT, SuperScript™ III, ImProm-II™, Omniscript, Sensiscript and Tth RT. At the end of the RT step the polymerase enzyme is heatactivated. It could also be that the reverse transcriptase polymerase and DNA polymerase is one and the same enzyme and that the enzyme only requires a DNA polymerase activation step after the RT step. An example of such an enzyme is Tth polymerase. This enzyme has both RNA-dependent reverse transcriptase and DNA-dependent polymerase activity. However, the active center of the DNA polymerase is covered by dedicated oligonucleotides, called aptamers. At temperatures below the optimal reaction temperature of the DNA-dependent polymerase component of Tth remains covered by the aptamers. At these temperatures the Tth enzyme only synthesizes a DNA copy of the RNA template. Once the reaction temperature is raised to 95 °C, the aptamers are removed and the DNA-dependent polymerase component will start to amplify the target sequence.

PCR amplification of the DNA target occurs in three steps: denaturation, annealing and extension. Each of these steps occur at a specific temperature for a fixed period of time. Denaturation is normally allowed to occur between 90 and 95 °C and leads to the dissociation of DNA strands. After this the reaction is cooled to between 40 and 70 °C to allow the primers to associate with their respective target sequences. This step is known as the annealing step and is primer specific. The temperature at which the primers anneal is critical. Too high temperatures would not allow the primers to associate with the DNA, resulting in no or poor amplification. Too low annealing temperature would ultimately lead to non-specific binding of the primers and non-specific amplification. Primers bound to the regions flanking the target DNA provide 3'-hydroxyl groups for DNA polymerase catalyzed extension. The most commonly used DNA polymerase is Taq, a thermo-stable enzyme isolated from the thermophilic bacterium, *Thermus aquaticus*. The DNA polymerase synthesizes new DNA strands along the template strands, using the primers as starting points. During the extension step the strands are amplified beyond the target DNA. This means that each newly synthesized strand of DNA will have a region complementary to a primer. There is an exponential increase in the amount of DNA produced as the three above mentioned steps are repeated in a cyclic fashion. In a traditional PCR these steps might be repeated 20 to 55 times. A problem, however, with PCR amplification is that the temperature required for DNA strand dissociation also results in DNA polymerase denaturation. This is partially overcome by the bioengineering of polymerases which are more thermal stable and have longer half-lives.

Even though RT-PCR is technically more difficult to perform and more expensive than ELISA, it has the ability to allow for the detection of low viral loads. RT-PCR is considered to be 102 to 105 fold more sensitive than traditional ELISA. RT-PCR also allows for the detection of several viral targets in the same reaction through the use of several primer combinations. This is called multiplexing. Although multiplexing is technically more demanding than a traditional simplex reaction it allows for a higher throughput in that a single sample can be tested for several viral strains in a single reaction. Primers used for multiplexing are chosen in such a manner that they result in amplicons of various sizes. This allows for post RT-PCR analysis using gel electrophoresis. Although RT-PCR saves time, allows for multiplexing and is more sensitive than ELISA, the reagents and instrumentation needed are expensive and require a higher level of technical expertise. Also, end product analysis using gel electrophoresis is laborious, relatively more expensive, time consuming and does not lend itself to automation. For these reasons the use of RT-PCR for routine screening is not feasible and has not replaced ELISA. It does, however, provide the industry with the opportunity to screen borderline cases, especially in the case of seed potato certification.

Quantitative PCR

In most traditional PCRs the resulting products are analyzed after the PCR has been completed. This is called end-point analysis and is normally qualitative of nature rather than being quantitative. For this sort of analysis, products are mostly analyzed on an agarose gel and visualized using ethidium bromide as a fluorescent dye. Direct correlation between signal strength and initial sample concentration is not possible using end-point analysis since PCR efficiency decreases as the reaction nears the plateau phase. Quantitative PCR, however, offers an accurate and rapid alternative to traditional PCR. Quantitative PCR offers the researcher the opportunity to amplify and analyze the product in a single tube using fluorescent dyes. This is known as homogeneous PCR. During a quantitative PCR the increase in fluorescence is correlated with the increase in product. Through the use of different specific, dyes quantitative PCR can be used to distinguish between different strains of a virus and even to detect point mutations. The major advantage of quantitative PCR is that analysis of resulting products using gel electrophoresis is not required. This means that quantitative PCR can be implemented as a high-throughput technique for sample screening.

Quantitative PCR has been described for detection and discrimination of PVY^{O} and PVY^{N} isolates and for reliable discrimination between PVY^{NTN} and PVY^{N} isolates.

Phyllody

Phyllody on a goldenrod (*Solidago* sp.).

Phyllody on a purple coneflower (*Echinacea purpurea*).

Phyllody is the abnormal development of floral parts into leafy structures. It is generally caused by phytoplasma or virus infections, though it may also be because of environmental factors that result in an imbalance in plant hormones. Phyllody causes the affected plant to become partially or entirely sterile, as it is unable to normally produce flowers.

The condition is also known as phyllomorphy or frondescence; though the latter may sometimes refer more generically to foliage, leafiness, or the process of leaf growth. Phyllody is usually differentiated from floral virescence, wherein the flowers merely turn green in color, but otherwise retain their normal structure. However, floral virescence and phyllody (along with witch's broom and other growth abnormalities), commonly occur together as symptoms of the same diseases. The term chloranthy is also often used for phyllody (particularly flowers exhibiting complete phyllody, such that it resembles leaf buds more than flowers), though in some cases it may refer to floral virescence.

Phyllody is characterized by the partial or complete replacement of floral organs with true leaves. Phyllody can affect bracts, the calyx (sepals), corolla (petals), the gynoecium (carpels/pistils), and the androecium (stamens). Phyllody may be partial, affecting only some sets of floral organs or even only half of a set of floral organs (e.g. only three petals out of six in a single flower); or it can be complete, with all the floral organs replaced by leaves.

Longitudinal section of a rose flower exhibiting phyllody. Despite the apparent hips, the reproductive organs are completely absent and have been replaced by leaves.

Longitudinal section of a normal developing rose hip.

Phyllody of the bracts is common among plants which bear catkin (amentaceous) inflorescences. They are very common among members of the genus *Plantago*, for example, as well as the common hop (*Humulus lupulus*). Involucral bracts of the flowers of members of the family Asteraceae like dahlias and dandelions, may also be affected.

Sepals that exhibit phyllody are usually hard to detect due to fact that most sepals already resemble leaves. Close examination, however, can reveal differences in venation in normal sepals and sepals that exhibit phyllody. The full development of perfect leaves from sepals is more common among flowers that have united sepals (monosepalous) than in flowers with separated sepals (polysepalous).

Phyllody of the petals can be expressed more mildly as a simple change in shape and color (in which case, it's more accurately virescence), or it can be expressed as fully formed leaves. It is more common among flowers which exhibit corollas of distinct petals (polypetalous) than in flowers in which the petals are fused into a single tube or bowl-like structure (monopetalous).

Phyllody of the stamens is rare. In fact, the stamens are the least likely of the floral organs to be affected by phyllody. This is thought to be because the stamens are the most highly differentiated organs in flowers.

In contrast, phyllody of the carpels is much more common than the corresponding changes in stamens. Usually, phyllody affects the proximal parts of the carpel (the ovary) more than the distal parts (the style and stigma). The ovule itself may be exposed on the edges or on the inner surface of the carpel if the ovary becomes leaf-like. If the ovule is affected by phyllody, it develops separately from the rest of the carpel. The best known example of phyllody of the carpels is found in the Japanese cherry (*Prunus serrulata*), in which one or both of the carpels can become leaf-like (although the distal half of the style and the stigma are usually unaffected). Incidentally, some Japanese cherry cultivars also exhibit "doubling" of the petals due to petalody, where a second corolla develops instead of stamens.

Causes

Biotic

In many cultivated plants, phyllody is caused by infections of plant pathogens and/or infestations of ectoparasites. Aside from exhibiting phyllody, they may also exhibit other symptoms like virescence, witch's brooms, chlorosis, and stunted growth. Examples of these biotic factors include:

- Phytoplasmas - Specialized prokaryotic microorganisms that cause more than 200 distinct plant diseases. They resemble other bacteria but lack cell walls and are filamentous or pleomorphic in form. They are obligate parasites of plant phloem tissue and are spread by insect vectors. They are the most common cause of phyllody. Evidence suggests that phytoplasmas downregulate a gene involved in petal formation, instead causing leaves or leaflike structures to form. Examples of commercially important phytoplasma diseases are aster yellows, apple proliferation, clover phyllody, and *Sesamum* phyllody.
- Viruses, like the rose rosette disease (RRD).
- Fungi, like the smut fungus *Sphacelotheca reiliana* of corn and the rust fungus *Atelocauda koae* which infects *Acacia koa.*
- Water molds, like *Sclerophthora macrospora* which infects more than 140 kinds of cereals including rice, corn, and wheat. The disease is more commonly known as "crazy top" because its most striking symptom is phyllody of the ears and tassels.
- Insect damage.

In addition to causing phyllody itself, insects and other ectoparasites also serve as disease vectors that can spread phyllody to other nearby plants. The most common of these insect vectors are leafhoppers, an example of which is *Hishimonus phycitis*, which transmits the phytoplasma-caused little leaf phyllody in eggplants. The broken-backed bug (*Taylorilygus apicalis*) is another insect vector of a phytoplasma-caused phyllody in species of *Parthenium*. Other ectoparasite vectors include eriophyid mites, like the rose leaf curl mite (*Phyllocoptes fructiplilus*) which is known to be the primary vector of the rose rosette disease; and the chrysanthemum rust mite (*Paraphytoptus chrysanthemi*) which transmits phytoplasma-caused phyllody in species of chrysanthemums.

Abiotic

Environmental abiotic factors like hot weather or water stress that result in an imbalance in plant hormones during flowering can cause phyllody. These can usually be differentiated from phyllody caused by biotic factors by the simultaneous presence of healthy and abnormal flowers. When conditions normalize, the plants resume normal flowering. The susceptibility of plants to environmentally caused phyllody can be genetic.

Artificial

Phyllody can be artificially induced by applying cytokinins (CK), plant hormones responsible for cell division, as well as apical dominance and axillary bud growth. Conversely, it can be subsequently suppressed with the application of gibberellins (GA), plant hormones responsible for stem elongation, flowering, and sex expression.

Barley Yellow Dwarf

Barley yellow dwarf is a plant disease caused by the barley yellow dwarf virus, and is the most widely distributed viral disease of cereals. It affects the economically important crop species barley, oats, wheat, maize, triticale and rice.

Biology

Barley yellow dwarf virus (BYDV) is a positive sense single-stranded RNA virus; the viron is not enveloped in a lipid coating. The virus is transmitted by aphids, and the taxonomy of the virus is based on genome organisation, serotype differences and on the primary aphid vector of each isolate.

The isolates and their major vectors (in parentheses) are:

- Subgroup I:
    - MAV, a less severe strain carried by aphids (grain aphid, Sitobion avenae),
    - SGV (Schizaphis graminum), and
    - PAV, a less severe strain carried by aphids (bird cherry-oat aphid, Rhopalosiphum padi, grain aphid, S. avenae, and others including rose-grain aphid, Metopolophium dirhodum).
- Subgroup II, called cereal yellow dwarf virus, however CYDV is now recognised as a separate species belonging to the genus *Polerovirus* of the family Luteoviridae:
    - RPV, the most severe strain carried by aphids (Bird cherry-oat aphid, Rhopalosiphum padi),
    - RMV (Rhopalosiphum maidis).

Pathology

Wheat plants dwarfed after infection with BYDV.

When aphids feed on the phloem of the leaf, the virus is transmitted to the phloem cells. Once inside the plant, the virus begins to replicate and assemble new virions. This process requires significant metabolic input from the plant, and causes the symptoms of barley yellow dwarf disease.

The symptoms of barley yellow dwarf vary with the affected crop cultivar, the age of the plant at the time of infection, the strain of the virus, and environmental conditions, and can be confused with other disease or physiological disorders. Symptoms appear approximately 14 days after infection.

Affected plants show a yellowing or reddening of leaves (on oats and some wheats), stunting, an upright posture of thickened stiff leaves, reduced root growth, delayed (or no) heading, and a reduction in yield. The heads of affected plants tend to remain erect and become black and discoloured during ripening due to colonization by saprotrophic fungi. Young plants are the most susceptible.

Infected wheat leaves have a reduced ability to photosynthesise.

The virus is transmitted from the phloem when the aphid feeds. When the aphid feeds, virions go to the aphid's hind gut, the coat protein of the virus is recognised by the hindgut epithelium, and the virion is allowed to pass into the insect's hemolymph, where it can remain indefinitely, but the virus cannot reproduce inside the aphid. The virus is actively transported into the accessory salivary gland to be released into salivary canals and ducts. The virus is then excreted in the aphid saliva during its next feeding.

The host range of BYDVs consists of more than 150 species in the Poaceae; a large number of grasses both annual and perennial are alternate hosts to BYVD and can serve as reservoirs of the virus.

There are two main sources by which a cereal crop might be infected:

- By non-migrant wingless aphids already present in the field and which colonise newly-emerging crops. This is known as "green-bridge transfer".
- By winged aphids migrating into crops from elsewhere. These then reproduce and the offspring spread to neighbouring plants.

Effect on Yield

This is variable since it depends on viral strain, time of infection and rate of spread. Most severe losses are from early infections and can be as high as 50%.

Control

"Green bridge" sources must be ploughed in as early as possible. Alternatively, a desiccant herbicide should be applied 10 days prior to cultivation. Insecticide sprays may be used at crop emergence.

Drilling dates prior to mid-October favors attacks from winged migrant aphids. However, yield penalties may be experienced from late drilling. Insecticide sprays in this instance are therefore aimed at killing the aphids before significant spread can occur.

OTHER COMMON PLANT DISEASES

Citrus Stubborn Disease

The Citrus stubborn disease is a plant disease affecting species in the genus *Citrus*. *Spiroplasma citri*, a Mollicute bacterium species, is the causative agent of the disease. It is present in the phloem of the affected plant and transmitted by several leafhoppers including *Circulifer tenellus* (beet leafhopper) and *Scaphytopius nitridus* in citrus-growing regions of California and Arizona and *Circulifer haematoceps* in the Mediterranean region.

The host most notably affected is sweet orange but the bacterium can also infect weeds such as periwinkle (*Vinca rosea*) and London rocket (*Sisymbrium irio*). Yellowed plants of Chinese cabbage and pak-choi (*Brassica rapa*) can be infected by *S. citri*. In the wild, shortpod mustard (*Hirschfeldia incana*) infested by the beet leafhopper, *Circulifer tenellus*, can prove to be an important reservoir of infection. *S. citri* can also be transmitted to China aster (*Callistephus chinensis*), Shasta daisy (*Leucanthemum × superbum*), red clover (*Trifolium pratense*) and radish (*Raphanus sativus*) by the leafhopper *Scaphytopius nitridus*. The bacterium has also been shown to experimentally infect white clover (*Trifolium repens*) using *Euscelis plebejus* as a vector.

Symptoms on citrus trees are variable but typically include small size with upright position. Fruits harvested from citrus trees with severe symptoms of citrus stubborn disease can be acorn-shaped or lopsided.

Hosts and Symptoms

As the name indicates, citrus stubborn disease affects citrus plants, most severely oranges (especially naval and mandarin varieties), grapefruit, and tangelo trees. Lemon and lime are also affected, but much less severely.

Symptoms of Citrus Stubborn disease are most prominent in immature plants but still appear on established trees. The primary symptom of Citrus Stubborn disease is the irregularity of fruit on the same tree. A tree with citrus stubborn disease will have fruits of differing sizes, shapes, and stages of maturity and typically lighter, smaller fruits than its healthy counterpart. Affected fruits will often drop prior to maturity and often have a characteristic acorn-like shape, which is easily seen by cutting the fruit in half. Coloration of the fruit is also affected. The blossom end remains green while the stem end is colored in affected fruits. Farmers would most readily use these symptoms as indicators that their crop may have been infected with *Spiroplasma citri*.

On leaves, *Spiroplasma citri* manifests itself as light mottling similar to that of nutrient deficiency and a more vertical orientation. Another indication of infection is bunchy growth caused by shortened internodes. Healthy plants will have comparably more outstretched branches and a rounded appearance. A tree with Citrus Stubborn Disease will have a very low yield and the fruit it does yield will not be comparable to a healthy fruit. Small leaves and upright, bunchy growth of branches is common in infected plants, as is dieback and leafdrop. However, Citrus stubborn does not affect canopy height, width, trunk diameter, and juice quality, regardless of the severity of the infection.

Diagnosis can be difficult because the disease is not often severe enough to produce outwardly apparent symptoms. In addition, when it is evident that a tree is diseased, the symptoms are very common and could be attributed to numerous other pathogens or environmental factors. To truly confirm the presence of *Spiroplasma citri* it must be detected by PCR or a tissue culture with spiroplasmas must be produced. It has been shown that PCR is the most effective diagnostic tool for citrus stubborn disease.

Management

The most effective way to prevent citrus stubborn disease is to prevent *Spiroplasma citri* from reaching and infecting young, susceptible plants. This is best achieved through a variety of cultural practices.

Spiroplasma citri is transmissible through several insect vectors, namely the beet leafhopper. One effective measure against the beet leafhopper is planting trap plants, such as sugar beets, that the insect vector favor but are not susceptible to citrus stubborn disease nearby, in attempt to draw the disease-carrying insects away from the citrus plants. To further increase the effectiveness a chemical component can be added by spraying insecticides on the trap plants, eliminating the insect vector and preventing the bacteria from reaching the citrus crop.

Older trees are much less susceptible to *Spiroplasma citri,* so it is most critical to be diligent in preventing infection while the tree is still maturing. Trees under 6 years old that have citrus stubborn disease should be completely removed, as they will never be productive, and infected trees older than 6 should be individually evaluated and either have symptomatic parts removed or be completely replaced with a healthy plant.

Citrus stubborn disease can be spread through grafting, so it is important to ensure that the mother tree is free of *Spiroplasma citri* before propagation. Also, trees should be obtained, if possible from areas where *Spiroplasma citri* is not viable to prevent bringing the pathogen into an orchard.

In addition, it is important to closely monitor the weeds in young orchards to ensure that they are not susceptible to *Spiroplasma citri*, and if susceptible ones appear, to remove them as soon as possible.

Downey Mildew

Downy mildew refers to any of several types of oomycete microbes that are obligate parasites of plants. Downy mildews exclusively belong to Peronosporaceae. In commercial agriculture, they are a particular problem for growers of crucifers, grapes and vegetables that grow on vines. The prime example is *Peronospora farinosa* featured in NCBI-Taxonomy and HYP3. This pathogen does not produce survival structures in the northern states of the United States, and overwinters as live mildew colonies in Gulf Coast states. It progresses northward with cucurbit production each spring. Yield loss associated with downy mildew is most likely related to soft rots that occur after plant canopies collapse and sunburn occurs on fruit. Cucurbit downy mildew only affects leaves of cucurbit plants.

Symptoms

Initial symptoms include large, angular or blocky, yellow areas visible on the upper surface. As lesions mature, they expand rapidly and turn brown. The under surface of infected leaves appears watersoaked. Upon closer inspection, a purple-brown mold becomes apparent. Small spores shaped like footballs can be observed among the mold with a 10x hand lens. In disease-favorable conditions (cool nights with long dew periods), downy mildew will spread rapidly, destroying leaf tissue without affecting stems or petioles.

Treatment and Management

Cultural Options

Because the downy mildew pathogen does not overwinter in midwestern fields, crop rotations and tillage practices do not affect disease development. The pathogen tends to become established in late summer. Therefore, planting early season varieties may further reduce the already minor threat posed by downy mildew.

Chemical Control

Fungicides applied specifically for downy mildew control may be unnecessary. Broad spectrum protectant fungicides such as chlorothalonil, mancozeb, and fixed copper are at least somewhat effective in protecting against downy mildew infection. Systemic fungicides are labeled for use against cucurbit downy mildew, but are recommended only after diagnosis of this disease has been confirmed. In the United States, the Environmental Protection Agency has approved oxathiapiprolin for use against downy mildew.

Organic Control

One way to control downy mildew is to eliminate moisture and humidity around the impacted plants. Watering from below, such as with a drip system, and improve air circulation through selective pruning. In enclosed environments, like in the house or in a greenhouse, reducing the humidity will help as well.

Plant Specific Mildews

Basil

Downy mildew of basil caused by *Peronospora belbahrii* has been a huge problem for both commercial producers and home growers. The disease was first reported in Italy in 2004, was reported in the U.S. in 2007 and 2008 and has been steadily increasing in prevalence, distribution, and economic importance since then.

Cucurbitaceae

Cucurbitaceae downy mildew (caused by *Pseudoperonospora cubensis*) is specific to cucurbits (e.g. cantaloupe (*Cucumis melo*), cucumber (*Cucumis sativus*), pumpkin, squash, watermelon (*Citrullus lanatus*) and other members of the gourd family). The disease is one of the most significant diseases of cucurbits worldwide.

Grapes

Plasmopara viticola is the causal agent of grapevine downy mildew.

Hops

Hop Downy Mildew (caused by *Pseudoperonospora humuli*) is specific to hops (*Humulus lupulus*). The disease is the single most devastating disease in Western United States hopyards, since the microbe thrives in moist climates. Infected young hop bines become stunted with thickened clusters of pale curled leaves. These spikes have a silvery upper surface, while the undersides of leaves become blackened with spores. These dwarfed spikes are called "basal spikes". 'Lateral' or 'terminal' spikes occur further up the vine. An entire hop crop could be devastated in only a few days.

Ornamentals

A new and particularly aggressive form of impatiens downy mildew has recently emerged as a major threat to the cultivation of ornamental impatiens in the United States, where they are one of the most popular ornamental plants.

Soybeans

Peronospora manshurica infects soybeans, reducing photosynthetic activity, yield, and quality. The fungus spreads by oospores on diseased leaves and/or on infected seed. The disease spreads in environments with high humidity and favors temperatures between 20-22 °C. Tufts of grayish to pale-colored sporangiophores on the underside of leaves easily distinguish the infection from other foliar diseases. The disease is often controlled using the fungicides mancozeb, maneb, or zineb.

Spinach

Downy mildew on spinach is caused by *Peronospora effusa,* an oomycete pathogen that poses a challenge to spinach production worldwide, especially in organic production.

Sunflowers

Plasmopara halstedii infects sunflowers, producing oospores which can remain dormant in the soil for many years.

Clubroot

Clubroot is a common disease of cabbages, broccoli, cauliflower, Brussels sprouts, radishes, turnips, stocks, wallflowers and other plants of the family Brassicaceae (Cruciferae). It is caused by *Plasmodiophora brassicae,* which was once considered a slime mold but is now put in the group Phytomyxea. It is the first phytomyxean for which the genome has been sequenced. It has as many as thirteen races. Gall formation or distortion takes place on latent roots and gives the shape of a club or spindle. In the cabbage such attacks on the roots cause undeveloped heads or a failure to head at all, followed often by decline in vigor or by death. It is an important disease, affecting an estimated 10% of the total cultured area worldwide.

Clubroot on cauliflower.

Historical reports of clubroot date back to the 13th century in Europe. In the late 19th century, a severe epidemic of clubroot destroyed large proportions of the cabbage crop in St. Petersburg. The Russian scientist Mikhail Woronin eventually identified the cause of clubroot as a "plasmodiophorous organism" in 1875, and gave it the name *Plasmodiophora brassicae*.

In 18th, 19th and early 20th century Britain clubroot was sometimes called *finger and toe*, *fingers and toes*, *anbury*, or *ambury*, these last two also meaning a soft tumor on a horse.

The potential of cultural practices to reduce crop losses due to clubroot is limited, and chemical treatments to control the disease are either banned due to environmental regulations or are not cost effective. Breeding of resistant cultivars therefore is a promising alternative.

In Cabbages

Cabbage clubroot.

Cabbage clubroot is a disease of Brassicaceae (mustard family or cabbage family) caused by the soil-borne *Plasmodiophora brassicae*. The disease first appears scattered in fields, but in successive seasons it will infect the entire field, reducing the yield significantly and sometimes resulting in no yield at all. Symptoms appear as yellowing, wilting, stunting, and galls on the roots. It is transmitted by contaminated transplants, animals, surface water runoff, contaminated equipment, and irrigation water. The pathogen can survive in a field for years as resting spores without a host present and will infect the next crop planted if it is a susceptible host. This pathogen prefers a wet climate and a pH around 5.7, so proper irrigation and the addition of compounds that raise the pH can be used to control this disease. Other control methods include sanitation to prevent transmission, chemical control, and resistant varieties.

Hosts and Symptoms

Cabbage Clubroot affects cabbage, Chinese cabbage, and Brussels sprouts most severely, but it has a range of hosts that it affects less severely like kohlrabi, kale, cauliflower, collards, broccoli, rutabaga, sea kale, turnips, and radishes.

Wilting and yellowing of plants in cabbage field.

Galls on plant roots.

Developing plants may not show any symptoms but as the plants get older they will start to show symptoms of chlorosis or yellowing, wilting during hot days, and exhibit stunted growth. Below ground, the roots experience cell proliferation due to increased auxin or growth hormone production from the plant as well as the pathogen. This causes the formation of galls that can grow big enough to restrict the xylem tissue inhibiting efficient water uptake by the plant. Galls appear like clubs or spindles on the roots. Eventually the roots will rot and the plant will die.

Disease Cycle

In the spring, resting spores in the soil germinate and produce zoospores. These zoospores swim through the moist soil and enter host plants through wounds or root hairs. A plasmodium is formed from the division of many amoeba-like cells. This plasmodium eventually divides and forms secondary zoospores that are once again released into the soil. The secondary infection by the zoospores can infect the first host or surrounding hosts. These secondary zoospores can be transmitted to other fields through farm machinery or water erosion. They form a secondary plasmodium that affects plant hormones to cause swelling in root cells. These cells eventually turn into galls or "clubs". The secondary plasmodium forms the overwintering resting spores which get released into the soil as the "clubs" rot and disintegrate. These resting spores can live in the soil for up to 20 years while they wait for a root tip to come in close proximity for them to infect.

Environment

Clubroot is a disease that prefers warmer temperatures and moist conditions. Ideal conditions for the proliferation of this disease would be a soil temperature between 20–24 °C and a pH less than 6.5; therefore, this disease tends to be prominent in lower fields where water tends to collect.

Management

Clubroot is very hard to control. The primary step for management and long-term control is exclusion of the disease. Good sanitation practice is important with regard to the use of tools and machinery in order to prevent the introduction of the pathogen to a disease-free field. It is not uncommon for an inattentive farmer or gardener to unknowingly carry in the pathogen after being previously exposed to it at a different time. One should avoid purchasing infected transplants of cabbage so as to prohibit the infestation of *P. brassicae*. Soil type is also an important factor in the development and spread of cabbage clubroot; the use of sand will allow for the plants to grow in well-drained soil, thereby eliminating the possibility of the pathogen to proliferate in a hospitable environment.

Although it is difficult to eradicate the pathogen once it is introduced to a field, there are several methods for its control. Keeping the soil at a slightly basic pH of 7.1–7.2 by the addition of agricultural lime as well as the integration of crop rotation will reduce the occurrence of cabbage clubroot in already infected fields. Fumigation using metam sodium in a field containing diseased cabbages is yet another way to decrease the buildup of the pathogen. Control and management practices on already infected fields help to reduce the overall impact that *P. brassicae* has on a field of cabbage and other cruciferous plants, but it is extremely difficult to rid an individual plant of the disease once it is already infected.

Cherry X Disease

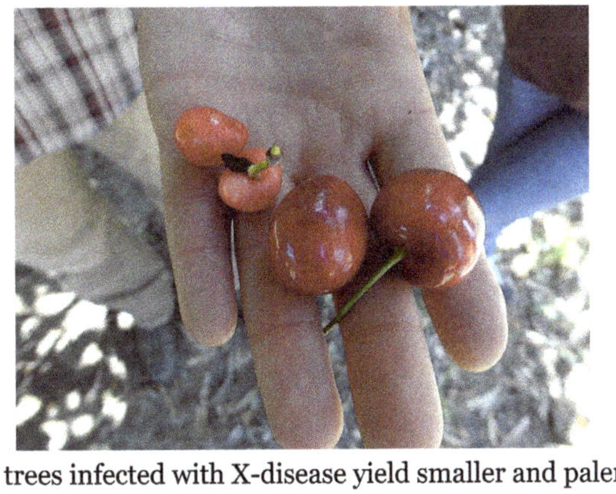

Cherry trees infected with X-disease yield smaller and paler fruit.

Cherry X disease also known as Cherry Buckskin disease is caused by a plant pathogenic phytoplasma. Phytoplasma's are obligate parasites of plants and insects. They are specialized bacteria, characterized by their lack of a cell wall, often transmitted through insects, and are responsible for large losses in crops, fruit trees, and ornamentals. The phytoplasma causing Cherry X disease has a fairly limited host range mostly of stone fruit trees. Hosts of the pathogen include sweet/sour cherries, choke cherry, peaches, nectarines, almonds, clover, and dandelion. Most commonly the pathogen is introduced into economical fruit orchards from wild choke cherry and herbaceous weed hosts. The pathogen is vectored by mountain and cherry leafhoppers. The mountain leafhopper vectors the pathogen from wild hosts to cherry orchards but does not feed on the other hosts. The cherry leafhopper which feeds on the infected cherry trees then becomes the next vector that transmits from cherry orchards to peach, nectarine, and other economic crops. Control of Cherry X disease is limited to controlling the spread, vectors, and weed hosts of the pathogen. Once the pathogen has infected a tree it is fatal and removal is necessary to stop it from becoming a reservoir for vectors.

Hosts

For Cherry X disease there are two types of hosts for the phytoplasma, reservoir and non-reservoir hosts. Reservoir hosts can survive for long periods while being infected with the disease. This allows them to be a constant food source for the leafhoppers which act to vector the phytoplasma from these hosts to other hosts in the area. Choke cherry is the most common reservoir host and a favorite food for the cherry leafhoppers. Other reservoir hosts include clovers and dandelions. Sweet/sour cherries, as well as almonds and Japanese plums are all fruit tree reservoir hosts for the Cherry X disease. All of these, once infected, can act as a source for the disease to be vectored from to other hosts. While non-cherry hosts can become infected they are not the preferred host of the phytoplasma. Because of the vectors preference for cherry trees, choke cherry which is a wild growing cherry species is the most common host of the disease. The range that Cherry X disease is distributed over is directly linked to the distribution of wild choke cherry populations.

Non-reservoir hosts are hosts that once infected do not allow for the disease to be spread. Peach and nectarine trees can be infected but they do not allow for the spread of the disease. This process which causes them to halt the spread of the pathogen is still not well understood. Peaches are commonly infected when near cherry orchards. Non-reservoir hosts are infected when cherry leafhoppers that are carrying the phytoplasma feed on non-reservoir hosts that are near a cherry orchard that has the pathogen.

Symptoms

The symptoms of Cherry X disease vary greatly depending on the host. On cherry hosts symptoms can usually first be seen on the fruits, causing them to be smaller in size with a leathery skin. Pale fruit is common at harvest time. It is common for symptoms to first be seen in a single branch. The branch may lose its older leaves, and the leaves tend to be smaller with a bronzed complexion.

The rootstock that the cherry is grafted onto can play a significant role in the disease symptoms seen. Rootstocks of Mahaleb cherry exhibit different symptoms from stocks of Colt, Mazzard, or Stockton Morello. When the scion is grafted onto Mahaleb, symptoms consistent with Phytophthora root rot can be seen. To distinguish between root rot and X-disease the wood under the bark at the graft union should be examined. If it is X-disease the wood at the union will have grooves and pits this causes a browning of the phloem and shows the cells in decline. This rapid decline is caused by the rootstock cells near the graft union dying in large quantities. Foliage begins to turn yellow and the curl upward and inward toward the leaf midrib. Trees infected with Mahaleb rootstock die by late summer or early the following year.

When Cherries are grafted onto Colt, Mazzard, or Stockton Morello rootstocks, there is a different range of symptoms. Affected leaves are smaller than normal and the foliage may be sparse. Dieback of shoot tips is common as the disease progresses. Fruit on branches are smaller, lighter, pointed, low sugar content, poor flavor, and a bitter taste.

Peaches are the next most common economic fruit host of the X-disease. Symptoms can be seen after about two months single branches will begin to show symptoms of their individual leaves. These leaves curl up and inward with irregular yellow to reddish-purple spots. These spots can drop out leaving "shotholes". Leaves that are affected by the disease will fall prematurely. After 2–3 years the entire tree will show symptoms.

Disease Cycle

Mountain leafhopper (Colladonus montanus) overwinters on winter annual weeds, particularly near streams and canals. Adults can be plentiful on sugarbeet during late winter/spring and migrate to favored weed hosts such as curly dock or burclovers in orchards. The Mountain leafhopper is most abundant vector found on cherry but does not reproduce on cherry. The mountain leafhopper (Colladonus montanus) spreads the disease from wild herbaceous hosts to woody hosts. It is believed that it is more responsible for the introduction of the disease into cherry trees, then in transferring them from cherry tree to cherry tree in an orchard. The cherry leafhopper (Fieberiella florii) reproduces on a broad range of woody hosts. The cherry leafhopper is more important in vectoring the disease from tree to tree within an orchard, since cherry is a favored host. After a leafhopper feeds on an infected host the pathogen has to undergo a latent period. During the latent period the pathogen spreads and multiplies inside the vector. Depending on temperature and the vector, the average latent period for the Cherry X disease is about a month or longer. The phytoplasma is then transmitted from the leafhopper to the tree when the leafhopper is feeding on the trees phloem. It's then spread throughout the tree becoming systemic. July through October is when the highest concentrations of pathogen are present in leaves of infected trees.

Environment

Leafhoppers are the only known vectors that can carry the X-disease from a wild host into peach and cherry orchards. Orchard trees are most often infected by insect vectors. In California where it was first noted, the two most important vectors were the mountain leafhopper, Collandonus montanus, and the cherry leafhopper, Fieberiella florii.

Mountain Leafhopper (Collandonus Montanus)

The mountain leafhopper survives on winter annual weeds during winter, usually near stream banks or canals. In late winter or spring, adults can be found in sugar beet fields and can then migrate to favored weed hosts (curly dock, burclovers) in orchards. The mountain leafhopper is most often the abundant vector found on cherry, however, cherry is not the preferred host and the leafhopper does not reproduce on cherry. Preferred hosts for the mountain leafhopper are; alfalfa, California burclover, clovers, curly dock, and sweet clovers. Of the preferred hosts alfalfa and curly dock cannot become infected with the disease itself but are just a host for the leafhopper. Occasional hosts are; vetches (in legume family) and sweet cherry. It's believed that the role of this leafhopper is introducing the disease into cherry orchards rather than spreading the disease between cherry trees within an orchard.

Cherry Leafhopper (Fieberiella Florii)

The cherry leafhopper has a more significant role in spreading the disease between cherry trees because cherry is a favored host. The leafhopper feeds and reproduces on a wide range of woody hosts. Preferred hosts for the cherry leafhopper are; box wood, lilac, myrtle, privet, pyracantha, sweet cherry, and viburnum. Of these preferred hosts only sweet cherry can become infected with the disease itself. Occasional hosts are; almond, apple and crabapple, apricot, bitter cherry, ceanothus, chokecherry, hawthorn, peach, pear, Japanese plum, and prune. Of these occasional hosts only chokecherry and bittercherry and occasionally almond, peach and Japanese plum can become infected with the disease itself.

There are seven known vectors that transmit the disease in western United States. These leafhoppers are *Colladonus geminatus, Fieberiella florii, Keonolla confluens, Scaphytopius delongi, Osbornellus borealis, Colladonus montanus,* and *Euscelidius variegatus*. Other possible leafhopper vectors are *Scaphytopius aculus, Paraphlepsius irroratus, Colladonus clitellarius,* and *Norvellina seminude.* Not a lot of information is available for ideal environmental conditions for the disease. However, conditions conducive to leafhoppers is most likely the key for the greatest spread of disease.

Management

There are numerous steps one has to take to try to manage the disease as best as possible. The aim is at prevention because once the pathogen reaches the cherry trees, disease will surely ensue and there is no cure or remedy to prevent the loss of fruit production as well as the ultimate death of the tree.

Pest Management

The first approach, which is the best approach at an effective management practice would be to eradicate or severely damage the Mountain and Cherry Leafhopper population because the leafhoppers are the number one vectors for this pathogen. To do this, pesticides (i.e. acephate, bifenthrin, cyfluthrin) could be applied or biological control (predators of the leafhopper) could be used. There should be a pre-season application of control measures as well as a post-season application. This is to maximize the effort at controlling both types of leafhoppers (Cherry and Mountain), thus cutting down the starting inoculum at both stages in the life cycle.

Weed Host Management

Some herbaceous hosts naturally have the Cherry X Disease. Once the spreads to the cherry hosts, with the help of the mountain leafhoppers, the cherry leafhoppers can spread the disease around to other woody hosts. Here are some approaches at management with each host type:

Herbaceous Hosts

The herbaceous hosts are common weeds (i.e. clovers, dandelions, alfalfa) that serve as a feeding ground for the mountain leafhoppers. The herbaceous hosts are the source of the X Disease, which is picked up and transmitted to the cherry hosts by the mountain leafhopper. For a control, conventional herbicides are effective. There exists a common herbaceous host, curly dock, which serves as the mountain leafhopper's main breeding ground. Getting rid of curly dock with an herbicide would be key to limit the population, thus limiting the spread of the X Disease to the cherry hosts.

Woody Hosts

After the disease moves on from the herbaceous host with the help of the mountain leafhoppers, it moves to the cherry hosts (i.e. bitter cherry and chokecherry). Once there, the infected trees should be destroyed and removed, along with all infected fruits. This is to prevent further spreading into other woody hosts such as peach, plum, apple etc. because once a tree is infected, it cannot be saved and it will become a source of the X Disease which the cherry leafhoppers can pick up and spread to the other woody hosts. In conclusion, all infected woody hosts should be removed and destroyed along with all infected fruits.

Cylindrocladium

Cylindrocladium diseases, caused by several species of Cylindrocladium (primarily C. scoparium, C. floridanum, and C. crotalariae), affect conifer and hardwood seedlings of many species. In the North Central and Northeastern States and in the Province of Ontario, species most commonly affected include red and eastern white pines, along with black and white spruces. In the Southern States, seedlings of black walnut, yellow-poplar, sweetgum, eucalyptus, and eastern white pine are the most commonly affected. In addition, cherrybark oak, northern red oak, dogwood, redbud, and several ornamental shrubs are susceptible.

Distribution

These diseases are known to occur in 20 Eastern and Central States and in Ontario, Canada. In addition, a species of Cylindrocladium has been isolated from soil samples in one nursery in Washington.

Damage

Cylindrocladium diseases, especially those affecting the roots, can cause significant seedling mortality. Sublethal infections can result in stunting, chlorosis, or top dieback so severe that many seedlings must be culled.

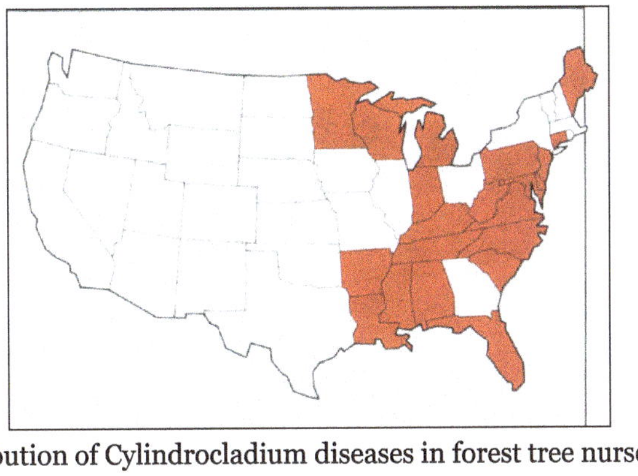

Distribution of Cylindrocladium diseases in forest tree nurseries.

Diagnosis

Infection by species of Cylindrocladium can result in a variety of symptoms. These include pre- and postemergence damping-off, root rot, foliage blight, and stem lesions.

Symptoms of root rot differ on seedlings of conifers and hardwoods. On conifers, look for necrosis of lateral and primary roots, frequently accompanied by blackening and slipping of the root cortex when the disease is at an advanced stage. On hardwoods, look for a pronounced blackening of the root cortex, frequently accompanied by longitudinal cracking. These symptoms occur on species such as yellow-poplar, black walnut, and sweetgum. Severe root infection can result in heavy mortality in both conifers and hardwoods.

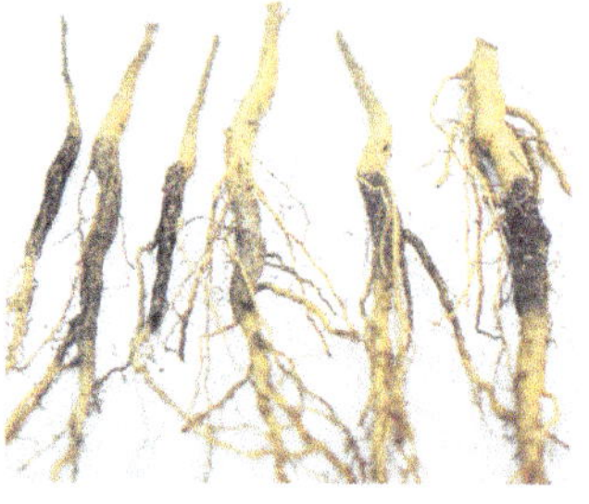

Blackened roots on yellow-poplar, a symptom typical of Cylindrocladium root rot.

Beds of black spruce severely affected by Cylindrocladium root rot.

Beds of black walnut infected by Cylindrocladium sp.

Stem infections on eucalyptus are frequently centered on leaf petioles. This fact suggests that infection begins in the foliage and progresses to the stems through the petioles.

Foliage blight symptoms on conifers such as eastern white pine are characterized by needle discoloration (yellowing and browning), necrosis, defoliation, and subsequent seedling mortality in severe cases.

When conditions are suitable, abundant spores (conidia) may be produced on infected plant parts, appearing as a white, powdery covering. Conidia of all species of Cylindrocladium are cylindrical with rounded ends. However, they vary in size and number of septations among species. Conidia of C. scoparium and C. floridan urn have one septum, and are 50-60 x 4.5-6.0 and 36-57 x 2.6-4.6 microns, respectively. Conidia of C. crotalariae have two or more septa, and are 58-107 x 4.8-7.1 microns. Species of Cylindrocladium also can be separated on the basis of vesicle shape. Vesicles of C. scoparium are primarily ellipsoid; those of the other two species are globose.

Symptoms of Cylindrocladium foliage blight (needle discoloration and necrosis) on 1-0 eastern white pine seedlings.

Stem of infected Euclayptus covered with white spore masses (conidia) of Cylindrocladium.

Conidiophores, conidia, and vesicles of C. scoparium.

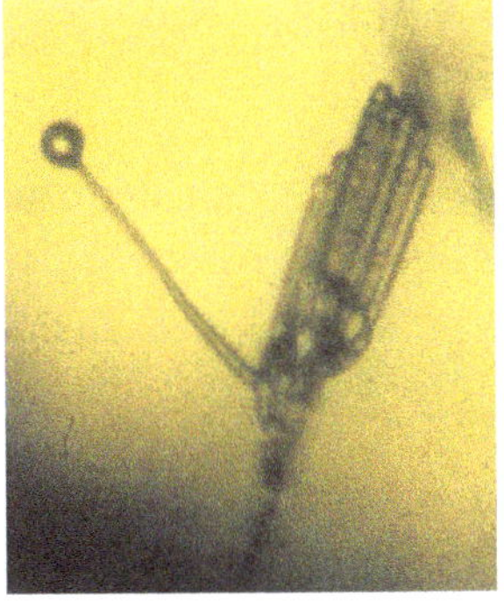

Conidiophores, conidia, and vesicles of C. floridanum.

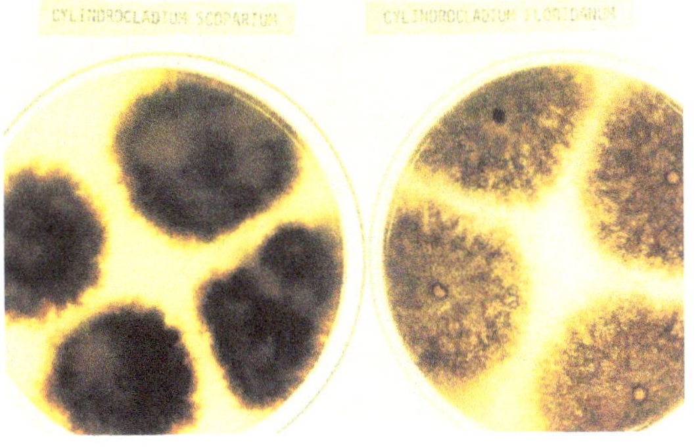

Reverse side of culture of Cylindrocladium sp. showing large numbers of microsclerotia.

Species of Cylindrocladium readily grow on most common laboratory media. Cultures are characterized by the production of abundant tiny, reddish-brown microsclerotia.

Biology

Cylindrocladium spp. survive and overwinter as microsclerotia in infected plant tissues and infested soil. When seedling roots come in contact with the microsclerotia, they germinate and infection occurs.

During periods of high humidity and rainfall, foliage and stem infection may also develop from infection by airborne conidia or ascospores. However, perithecia and ascospores are rarely produced in bareroot or container nurseries and have only been observed in association with C. crotalariae infections.

The Cylindrocladium spp. also have the unique characteristic of tolerating a wide pH range for fungus growth and host infection. This reduces the effectiveness of nursery cultural control techniques.

Control

- Prevention - Early detection, diagnosis, and evaluation of damage are essential. Delineate and avoid infested nursery sites as much as possible. Avoid nursery site contaminations through movement of either infected seedlings or infested soil between and within nurseries.

- Cultural - Favor nonhost nursery cover crops such as corn and grasses over known host cover crops such as soybeans, clover, and alfalfa. Maintain optimum nursery bed seedling densities to promote seedling aeration and growth and to reduce damage to both roots and foliage. Rogue nursery seedbeds and cull lifted material. Remove and destroy all seedlings with either discolored or wilted foliage. Cull those seedlings with 25 percent or more visible root rot damage to the taproot.

  Particularly for hardwood seedlings, minimize storage periods and maintain cold room storage temperatures at 35 to 40 °F.

- Chemical - In infested nurseries, fumigate seedbeds immediately before sowing. Special considerations may be needed for soil fumigation preceding the cover crop to provide adequate endomycorrhizae on hardwood seedlings. The most effective soil fumigant is a formulation of 67 percent methyl bromide and 33 percent chloropicrin. Employ deep soil fumigation (minimum of 12 in) when growing highly susceptible and deep-rooted hardwood species such as black walnut, yellow-poplar, and sweetgum.

- Apply foliar fungicides to prevent foliage and shoot diseases on susceptible conifer and hardwood seedlings. Benomyl and chlorothalonil have been effective for this purpose on eucalyptus seedlings in a south Florida nursery; however, special local need registrations may be required in other States.

- Dip seedling roots before transplanting. Dipping roots in solutions of benomyl has been effective in reducing losses from Cylindrocladium root rot on transplanted conifer seedlings in the North Central and Eastern States.

References

- Bacterial-leaf-spot, plant-disease, pest-problem-solver: planetnatural.com, Retrieved 8 January, 2019
- Botrytis fruit rot / gray mold on strawberry | nc state extension publications". Content.ces.ncsu.edu. Retrieved 2017-12-11
- Bhattacharya, shaoni (2017-02-09). "deadly new wheat disease threatens europe's crops". Nature. 542 (7640): 145–146. Bibcode:2017natur.542..145b. Doi:10.1038/nature.2017.21424. Pmid 28179687
- Chemical control of powdery mildew of apple in warmer climates of himachal pradesh, india". Actahort.org. Retrieved 2018-04-24
- Petanović, r.; kielkiewicz, m. (2009). "plant–eriophyoid mite interactions: specific and unspecific morphological alterations. Part ii". Experimental and applied acarology. 51: 81. Doi:10.1007/s10493-009-9328-1
- Dickinson, matt. "moble units of dna in phytoplasma genomes". Molecular microbiology. Consumer health complete - ebscohost. Retrieved 12 april 2013
- George n. Agrios (2005). Plant pathology (5th ed.). Burlington, ma: academic press. Isbn 978-0-12-044565-3

5

Diagnosis and Control of Plant Diseases

The diagnosis of plant diseases refers to the process of determining which disease is affecting the health of the plant. Some of the principles of plant disease control are avoidance, exclusion, protection and eradication. All these diverse principles related to diagnosis and control of plant diseases have been carefully analyzed in this chapter.

PLANT DISEASE DIAGNOSIS

Let us discuss the various steps/activities which are associated with accurate plant disease diagnosis. The process may vary with different diseases and conditions but the overall process is relatively consistent. The steps all require careful observations and questions. The steps include:

Proper plant identification. Identification of affected plants is one of the first steps in diagnosing a plant disease. Both scientific and common names of the plant should be noted. Common names should not be relied upon since some distinctly different plant species may have the same common name, and the common name used in one area may be used for a completely different species in another area. The common name “vinca” has been used to describe plants belonging to two different genera, Vinca, a perennial, and Catharanthus, an annual. Another example is “monkey grass” which is used to describe Liriope and Ophiopogon (mondo grass). An example from forestry is “cedar” which is used to describe eastern red cedar (Juniperus), western red cedar (Thuja), Port Orford cedar (Chamaecyparis), incense cedar (Libocedrus), and Atlas cedar (Cedrus). Obviously the use of common names can cause confusion in identification and recognition of problems.

In addition to knowing the common and scientific names of an affected plant, it is important to know the specific variety or cultivar, whenever possible. A great variation in susceptibility to a specific disease may occur within different cultivars of a plant species. For example, when we look at the susceptibility of wheat to wheat stem rust caused by Puccinia graminis f. sp. tritici, we know that all wheat cultivars are not susceptible to all races of P. graminis. The major control measure for this disease is based on planting wheat cultivars each year that are resistant to the pathogen races that are predicted to be present during the growing season. Tomato cultivars having the “Better Boy” genetic background are generally resistant to root-knot nematodes while those with the genetic background of the variety “Rutgers” are susceptible, so knowing the genetic background of a cultivar can be important. Knowing the cultivar and its susceptibility to various diseases can narrow down the possible disease agents to consider.

Knowing the identity of the plant species affected allows the pathologist to utilize various resources that contain lists of plant diseases associated with specific plants. These lists are very helpful in suggesting possible pathogenic agents.

Recognize healthy plant appearance. It is important to know the normal appearance of the plant species you are investigating. Each plant species has special growth habits, colors and growth rates. If you do not know what to expect of the plant you cannot recognize when something is wrong. Does the plant normally have new foliage that is yellow or red and becomes darker green as the foliage ages? Many ornamental shrubs have been developed and marketed for the ornamental value of such brightly colored new growth. These plants are highly prized for this coloration; however, if an individual does not know that this coloration is the normal appearance of the plant, s/he may think that the plant is diseased. It is important to know what the normal appearance of a plant is before you decide there is a problem. It is also important to remember that appearance can vary with different cultivars. Some plant cultivars have naturally yellow to pale green leaves (e. g. new hosta cultivars, herbs like golden oregano, and coleus varieties) which at first glance appear to have symptoms of under-fertilization, root stress or soil pH problems.

Once the "normal" appearance of the specific plant is determined, several comparisons can be made between the problem plants and healthy plants. Compare characteristics such as overall size, shape, and coloration; leaf shape, size, coloration, and distribution; root distribution and coloration; and bark, stem or trunk texture and coloration. It is also important to note normal events, such as leaf drop, that may occur in a healthy plant. For example, some holly species normally drop leaves in the spring.

The affected parts of the plants should also be noted. Are there symptoms on the roots, leaves, stems, flowers, or fruit? Is the entire plant involved? Is only one limb or side of a plant involved? Answers to these questions can assist in the identification of the problem.

Symptoms and Signs

Identify characteristic symptoms. Describing the characteristic symptoms exhibited by a specimen can be very difficult to do accurately. Because of this, it is often difficult, if not impossible, to determine what is wrong with a plant when a person is describing symptoms over the phone. As a test of this you may want to take a plant exhibiting symptoms and have three different individuals describe the symptoms that they observe on a sheet of paper. Next, compare the descriptions. Do the descriptions vary significantly? Could you visualize the symptoms by the way any one of the individuals described the diseased plant? Symptoms can often be grouped as follows:

- Underdevelopment of tissues or organs. Examples include such symptoms as stunting of plants, shortened internodes, inadequate development of roots, malformation of leaves, inadequate production of chlorophyll and other pigments, and failure of fruits and flowers to develop.
- Overdevelopment of tissues or organs. Examples include: galls on roots, stems, or leaves, witches' brooms, and profuse flowering.

- Necrosis or death of plant parts. These may be some of the most noticeable symptoms, especially when they affect the entire plant, such as wilts or diebacks. Other examples include shoot or leaf blights, leaf spots, and fruit rots.
- Alteration of normal appearance. Examples include mosaic patterns of light and dark green on leaves, and altered coloration in leaves and flowers.

Diseases also involve a progression of symptoms that can vary significantly. The progression of symptoms is one of the most important characteristics associated with problems caused by biotic agents. Diseases can result in primary and secondary symptoms. For example, decayed roots on a tree may be a primary symptom while the toppling over of the tree or windthrow is a secondary symptom. At later stages of a disease, secondary invaders may also obscure the original disease symptoms so that symptoms observed at the later stages of the disease are not typical of the symptoms developed in response to the original pathogen.

It is important to look for a progression of disease symptoms in plants exhibiting problems. In some cases, such as improper herbicide usage, symptoms observed may be similar to spots present as a result of an infectious agent. The difference is that with herbicide injury, the symptoms usually appear suddenly and there is no observable progression of symptoms. The spots may also follow spray patterns of the herbicide. Herbicides, such as 2,4-D, can cause leaf distortion which may be confused with viral diseases. However, when new leaves form, they will generally be free of symptoms, indicating a lack of symptom progression.

Identify symptom variability. Variations in symptoms expressed by diseased plants may lead to an improper diagnosis. These variations can result from a couple of factors. It is possible that there is more than one problem present, and in some cases there may be more than one pathogen infecting a plant. Symptoms associated with these infected plants may be significantly different from the symptoms expressed in response to each of the different pathogens acting separately. The disease symptoms exhibited by multiple pathogens infecting a plant may be either more severe or less severe than if the plant were infected with just one of the pathogens. This is commonly seen in multiple infections due to viruses. An example of this is shown in Figure which shows peach seedlings infected with single or multiple viruses. The seedling on the left is infected with both Prune dwarf virus and Prunus necrotic ringspot virus. The seedling in the middle is infected with Prune dwarf virus alone and the seedling on the right is infected with Prunus necrotic ringspot virus alone.

Figure: Peach seedlings infected with various viruses alone or in combination. Peach seedlings infected with both Prune dwarf virus and Prunus necrotic ringspot virus (seedling on left), infected with Prune dwarf virus (seedling in middle) and infected with Prunus necrotic ringspot virus (seedling on right).

Look for signs of biotic causal agents. Signs of plant disease agents are the observable evidence of the actual disease-causing agent. Signs may include the mycelia of a fungal agent, fungal spores, and spore-producing bodies. Indications of insects causing problems may include the actual insect, insect frass, mite webbing, and insect eggs. Signs are much more specific to disease-causing agents than are symptoms and are extremely useful in the diagnosis of a disease and identification of the agent causing the disease. The use of a hand lens and a knife can be valuable for a diagnostician in the field. Cutting into the bark of ornamental plants and trees at the soil surface may lead to the observation of mycelial mats of root rot fungi such as Armillaria spp. Bacterial ooze can be observed by cutting stems and placing them in water. Masses of different spores such as rust spores on leaves can also be important in disease diagnosis. Powdery mildews are typically diagnosed by the observation of the gray to white mycelia and conidia observed on the surface of leaves and flowers.

White mycelia of Armillaria growing under bark of peach tree.

Bacterial ooze from cut tomato stem infected with Ralstonia solanacearum.

Stem rust on barley. caused by Puccinia graminis.

Powdery mildew on apple blossom cluster caused by Podospaera leucotricha.

Dissecting and compound microscopes are useful for the observation of specific spores and spore structures, and can lead to further identification of possible disease agents. Knowledge concerning the use of microscopes and a hand lens is vitally important to the diagnostician. Signs of plant disease agents can often be overlooked unless careful observations are conducted. Signs are not visible when taking a quick ride by plants looking through the windshield of a truck and may not even be visible to the naked eye.

Plant Parts Affected

It is important to note if the symptoms observed are associated with specific plant parts. For example, is a wilt observed correlated with a disruption of the vascular system which may be indicated by browning of the vascular system or are the roots of the plants abnormal including rots, decreased feeder roots, etc.; are necrotic lesions observed strictly on younger leaves? The symptoms of some diseases are most commonly seen on specific plant parts and this observation can be important in diagnosis.

Observe Patterns

Check distribution of symptoms. One of the first things that a diagnostician should note is how the diseased plants are distributed over the affected area. Are they distributed uniformly across an area or are they localized? Is there a definite pattern to the distribution? For example, does it occur only along the edges of a greenhouse near open windows, next to roadways or driveways, in low spots of a field, along a planted row, or is it affecting plants at random in a field? This distribution can be especially important in looking at the possibility of non-infectious problems, such as improper herbicide use or various soil factors. A uniform pattern on an individual plant and uniform damage patterns over a large area are generally not associated with biotic agents, but are usually due to abiotic agents.

How prevalent is the problem? Are all plants affected? Infectious problems generally occur over time and there is a progression of symptoms. Rarely will all of the plants be affected. Generally, disease problems caused by biotic agents will be observed when they are causing problems on a low percentage of plants at least at the start of the disease, unless there were extenuating circumstances, such as the use of infected seeds. Even then, rarely will 100% infection be observed. When a problem appears in 100% of the plants, it more commonly results from factors such as soil conditions (deficiencies or toxicities), adverse climatic factors (cold temperatures, hail, drought, etc.), or toxic chemicals (improper pesticide use, growth regulators, air pollutants, such as ozone, etc.).

What has been the progression of symptoms on plants in the affected area? If the symptoms all appeared at the same time and there has been no further development of symptoms, this would indicate a possible episodic event such as a change in temperature or possible improper chemical usage. If however, the symptoms started in one area and slowly spread to other areas and the severity of disease symptoms changed over time, this would be more indicative of the presence of a biotic agent. Biotic agents can also include insects and mammals, such as voles, which may be feeding on plants in an area.

Check for host specificity. Is the problem occurring in only one plant species or are different plant species affected? If different plant species are affected, this suggests the possibility of a non-infectious problem which could be related to cultural or environmental problems. However, Phytophthora and Pythium root rots can cause problems on many different plant species; therefore, the fact that more than one plant species is affected does not completely eliminate infectious agents. If there is more than one species of plant involved, are these plants closely related and can they be infected by a common pathogen?

Review the cultural practices and growing environment. It is vital that a diagnostician question the activities that have been conducted around the affected plants. The problem may not be due to anything that the grower has done; the problem could be related to what his/her neighbor has done. Information pertaining to the growing environment to which the affected plant has been exposed is a vital piece of the puzzle. It is especially important to document changes in the environment.

Environmental factors to consider include: extreme temperatures (freezing and heat), rainfall, hail, lightning, prolonged drought, temperature inversions (important in possible air pollutant damage and pesticide drift) and prevailing winds. All of these abiotic factors can be important to the problem. Site factors such as soil type, possible drainage problems, and soil pH should also be evaluated.

Cultural and maintenance activities can be significant. What pesticides or other chemicals have been applied? At what rate and when were they applied? Who applied the chemicals? What equipment was used in its application? What other activities have occurred? Has someone been mowing in the area? Has the highway department been working along the roadway, possibly applying herbicides? Have any unusual occurrences or weather patterns been noted? Many times careful investigation by the diagnostician is required because, in some cases, someone may have done something improperly and may be unwilling to admit their error.

Laboratory Tests

Sometimes neither symptoms nor signs provide enough specific or characteristic information to decide the cause of an infectious plant disease. In such cases, it may be necessary to bring a sample back to the laboratory for further tests to isolate and identify the causal agent. This can be a time-consuming and labor-intensive process that takes specialized skills.

Incubation of plant material. One of the first steps when getting back to the laboratory may be to place a sample of the diseased tissue under conditions that will allow an infectious agent to grow and possibly induce sporulation. This can be accomplished by placing a leaf in a moist chamber. A moist chamber can be a sterile petri dish containing a wet filter paper in the bottom of the dish and a triangle of glass tubing on which the sample is placed so that the sample is not directly on the wet filter paper but is exposed to humid conditions. This type of moist chamber will work for small and relatively flat specimens such as leaves. Plastic bags or boxes may be necessary for larger specimens. Saprophytes that are present on the specimen can also be encouraged to grow in a moist chamber and a brief surface swab with 70% isopropanol or 0.1-1% sodium hypochlorite may be useful in reducing these saprophytes. Moist chambers are generally incubated at room temperature.

Isolation and identification of biotic plant disease causal agents. Isolation of fungi usually requires that pieces of infected plant tissue be placed on various nutrient media. The organism that grows out of this tissue is then isolated in pure culture. Bacteria are often isolated by chopping up infected tissue in a small amount of sterile water. This water:bacteria suspension is then streaked onto a bacteriological medium such as nutrient agar. Several problems can occur when trying to isolate the plant pathogenic agent. The infected plant tissue may contain one or more saprophytes which have moved into the infected tissue. These saprophytes may outgrow the plant pathogen on the nutrient medium, obstructing accurate identification of the pathogen. In some cases where a specific plant pathogen is suspected, a medium selective for the suspected pathogen may be utilized. It is also beneficial to attempt to isolate the plant pathogen from the margins of the diseased tissue where the pathogen is more numerous or more active than saprophytes that quickly colonize the recently killed tissue.

Once an organism is isolated, is that organism the true cause of the problem? Conducting Koch's postulates, which involves the inoculation of healthy plants, may be necessary to conclusively answer this question, especially if the organism has not been previously reported as a plant pathogen on that host. Koch's postulates are seldom conducted for routine diagnoses, but may be extremely important for new diseases and for research. Inoculation of a healthy host and obtaining the symptoms originally observed in the field may be difficult. This may be due to problems in replicating the conditions through which the host was inoculated and also in reproducing the environmental conditions present when the host became infected. It is often impossible to replicate in the laboratory the original conditions present during the disease development.

Once an infectious plant pathogen is successfully isolated, the organism must be identified. There are estimated to be some 1.6 million fungal species, most of which are not infectious pathogens. Many fungi and bacteria have never been isolated and identified. The characteristics upon which their identification is based are often complex and specialized training is necessary to be able to identify these fungi and bacteria. Diagnosticians with experience are often able to identify the most commonly isolated organisms. Identification of plant pathogenic nematodes also requires a trained individual.

Diagnostic tests for identification of biotic causal agents. A major problem in identification of biotic causal agents is the inability of some infectious pathogens to grow on artificial media. Viruses, as well as some fungi (e.g. powdery and downy mildew causing agents) and some prokaryotes (e.g. phytoplasmas), require a living host in order to grow. In cases where the plant pathogen is difficult or impossible to grow on artificial media, other methods may be used for their detection, such as the use of serological tests for viruses. Viral identification is often accomplished utilizing ELISA (enzyme-linked immunosorbent assay) which is based on the binding of an antibody produced to a specific virus with the virus in the infected plant material. More tests are currently being developed using the polymerase chain reaction (PCR) for detection of specific organisms. These types of reactions take specialized equipment and reagents, and the tests are not commonly done outside diagnostic and research laboratories. Other techniques used for the identification of viruses include negative staining and electron microscopy to view the viral particles in plant tissue or suspensions.

PCR and ELISA tests, as well as other laboratory tests, may be used for organisms that will grow on artificial media. Additional tests may include analysis of fatty acids of organisms, carbohydrate utilization (i.e. BIOLOG test), and enzyme activity testing (i.e. pectinase, isozyme patterns).

Diagnostic tests for identification of abiotic plant disease causal agents. It is extremely important to look for abiotic factors that may be important in observed symptoms. Soil and water tests may be necessary to determine pH, nutrient composition, salinity, and other factors such as pesticide residues that may induce various symptoms. It may also be important to get samples of plant tissue analyzed for nutrient content to determine if there are macro- or micronutrient deficiencies or toxicities.

Final Diagnosis

Diagnosis is a form of hypothesis testing, where the hypothesis is simply the identity of the disease, and a good diagnostician goes through multiple iterations of the scientific method (seeking evidence through testing that supports or refutes the hypothesis that s/he generates). These hypotheses are generated through observations of the plant, environment, and information from the grower. When all of the information is successfully collected, literature sources should be consulted to determine what is already known about diseases and disease-causing agents associated with the identified plant. Information can be obtained from published resources including plant disease compendia, plant disease indexes, technical notes, commodity newsletters, online resources, and personal communication with plant disease experts. When no information is available on the specific plant, information on diseases and disease causing-agents of similar plants may be useful. There may also be rare cases where no information is available related to the disease. Then, extensive testing may be necessary to determine the identification of the plant pathogen. When this type of testing is required, it may take a long time to develop research-based control recommendations and control measures may have to be based on diseases of similar etiology. If these diseases have occurred in other areas of the world, control measures that have been previously developed in other areas may be useful.

The student should keep in mind that s/he is a detective. Causal agent identification and diagnosis of plant problems is just like a detective investigating an assault or murder case, only in this case, the victim is a plant. All clues should be investigated. Some clues may lead down blind alleys while others will lead down the correct path. It is important to note that there are exceptions that exist and these exceptions must be considered. It is the compilation of the information and clues that will ultimately lead to the most accurate diagnosis.

MORPHOLOGICAL SYMPTOMS OF PLANT DISEASES

Thousands of plant diseases have been recorded throughout the world, many of these causing heavy crop losses. Early detection and accurate diagnosis is essential for the effective management of plant disease. Thus the first step in studying any disease is its timely detection of the diseased plant. Quick initial detection is largely based on the signs and symptoms of disease.

Signs are the visible physical presence of either the pathogen itself or the structures formed by the pathogen. Common examples of easily detected signs are those such as the fungal mycelia and spore masses of downy mildews observed on infected leaves and the bacterial ooze of *Xanthomonas* leaf streak disease on rice.

Symptoms are the visible changes that occur in the host plant in response to infection by pathogens. For any disease in a given plant, there is the characteristic expression of symptoms, usually occurring in a sequential series during the course of the disease. This series of symptoms depicting the disease picture is referred to as the disease syndrome.

Morphological symptoms may be exhibited by the entire plant or by any organ of the plant. These have been categorized into different groups for easy of study. Primarily, morphological symptoms of plant diseases can be categorized into 6 different types.

- Necroses
- Growth abnormalities

- Metaplastic symptoms
- Proleptic symptoms
- Color changes
- Wilts

Necroses

Necroses are caused due to necrosis or death of plant cells. The affected plant tissue usually turns brown to black in color. Necrotic symptoms could appear in any part of the plant such as in storage organs, in green tissues, or in woody tissues.

Necrosis in Storage Organs

Death of cells in storage organs terminates in decomposition or decay referred to as a rot. Two types of rots are identified as Dry rot and Wet rot on storage tissues.

Soft rots are those where the pathogen breaks down the host cell walls, resulting in the exudation of juices from the infected tissue. The organ becomes mushy or pulpy and a foul smell often develops due to colonization by secondary invaders. Many fungi and bacteria cause soft rots on several fruits and vegetables. Species of the fungus, *Rhizopus* and bacterium *Erwinia* are two such commonly found pathogens causing soft rots.

In a dry rot, the storage organ becomes hard and dry, and in some diseases, there is rapid loss of water and the infected organs become shriveled, wrinkled, and leathery. Dry rots showing such symptoms are referred to as mummifications.

Necrosis in Green Tissues

Necroses on green tissue are termed differently based on the nature of symptoms and the type of green tissue. The term, damping off refers to the sudden wilting and topping over of seedlings as a result of extensive necrosis of tender tissue of the roots and stem near the soil line, due to the attack of soil-borne pathogens such as fungus, *Pythium*. This fungus is known to cause damping off in an assortment of seedlings such as that of brinjal, chilli, mung beans, tobacco, tomato, and *Cucurbita*.

Necrosis, (the affected plant tissue).

A spot refers to a well-defined area of gray or brown necrotic tissue. Spots are very common on leaves and fruits and are probably the most familiar necrotic symptom. Sometimes the necrotic tissue within a leaf spot may crack and fall off from the surrounding green tissue leaving an empty space. Such a symptom is known as a shot hole. Minute or very small spots are sometimes referred to as flecks or specks.

When dark mycelia of a fungal pathogen appear on the surface of necrotic spot, blotting the leaves, shoots, an stems as large and irregular spots, the symptom is referred to as a blotch.

Both streaks and stripes occur in grasses and are elongated areas having dead cells. Streaks occur along the stem and veins, while stripes are in the laminar tissues between veins. Net necrosis is a symptom resulting from an irregular pattern of anastomoses between streaks or stripes.

Blights are characterized by the rapid death of entire leaves including the veins or parts of the leaves. Blights also could occur on flowers and stems. Scorches resemble blights, but there necrosis occurs in irregular patterns between veins and along leaf margins.

Firing is sudden drying, collapse and death of entire leaves. Firing occurs in response to the activity of root rot and vascular wilt pathogens.

Scald is the blanching of epidermal and adjacent tissues of fruits and occasionally of leaves. The sudden death of unopened buds or inflorescence is referred to as blast. Extensive necrosis of fruits that resemble in premature dropping is called shelling.

Necrosis in Woody Tissues

Necrosis of woody tissue often brings about various types of die-back symptoms. Dieback is the extensive necrosis of a shoot from its tip downwards.

Restricted necrosis of the bark and cortical tissue of stems and roots is termed as a canker. In cankers, necrotic tissue in the sunken lesions is sharply limited, usually by a callus from adjacent healthy tissue.

When woody tissues are diseased, they may exude different kinds of substances. When the exudate is gummy, the symptom is called gummosis, while it is resinosis when it is resinous. If the exudate is neither gummy nor resinous, it is described as bleeding.

Abnormalities in Growth

Many disease symptoms are associated with growth changes in diseased plants. These could be caused by either reduced growth due to hypoplasia and atrophy or excessive growth due to hyperplasia and hypertrophy.

Hypoplasia and Atrophy

- Hypoplasia is the failure of plants or plant organs to develop fully due to a decreased production of the number of cells. Hypoplasia results in plants or plant parts of sub-normal size.
- Atrophy is the reduction in the size of plant cells produced. This also results in stunted plants or plant parts.
- Dwarfing is the failure of a plant or a plant part to attain its full size.

- Rosetting is a condition where the internode of a plant do not elongate, and hence, the leaves appear close together in a cluster.

Hyperplasia and Hypertrophy

- Hyperplasia is the enlargement of a plant tissue due to excessive increase in the number of plant cells produced. Hyperplasia results in overdevelopment in size of plants or plant organs.
- Hypertrophy is excessive growth due to the enlargement of individual cells. This condition also results in the overdevelopment in size of plants or plant organs.
- Hyperplasia and hypertrophy could result in the enlargement of leaves and fruits, and the enlargement of stems and roots.

Enlargement of Leaves and Fruits

Several symptoms expressing enlargement of leaves and fruits are commonly observed among diseased plants.

Curling which is the bending of the shoot or the rolling of the leaf, is a result of over-growth on one side of an organ. Often viral diseases cause such leaf distortions due to irregular growth of the lamina. Extreme reduction of the leaf lamina brings about the symptom known as the Shoe-string effect.

The puckering or crinkling of leaves due to different growth rates in adjacent tissue is known as savoying. Overgrowth of epidermal and underlying tissues of leaves, stems, fruits and tubers may result scab formation. Scab consists of raised, rough, and discrete lesions. These are often sunken and cracked, giving a typical scabby appearance.

Localized swellings or enlargement of epidermal cells due to excessive accumulation of water is termed intermuscence and the diagnostic symptom is the appearance of a blister.

Enlargement of Stems and Roots

Symptoms causing enlargement of stems and roots are termed differently based on their nature. Excessive accumulation of food material in stems, above a constricted area produces a swelling termed sarcody.

Localized swellings that involve entire organs are termed tumefaction. Commonly exhibited tumefactions are galls, clubs, and knots.

Excessive development of adventitious organs results in fasciculation, that is the clustering of organs around a focal point. Such examples include witch's broom and hairy root. Witch's broom is a broom-like mass proliferation due to the dense clustering of branches of woody plants while hairy root results due to excessive development of roots.

Fasciation is the broadening or flattening of cylindrical organs such as stems. The continued development of any organ after it has reached a stage beyond which it normally does not grow is known as proliferation. The outgrowth of tissue in response to wounding is known as a callus. Callus formation is found to form around most cankers.

Metaplastic Symptoms

Metaplastic symptoms are those which form when tissues change from one form to another. Such symptoms include phyllody, the development of floral organs into leaf-like structures, juvenillody, the development of juvenile seedlings on mature plants and russeting, a superficial browning of surfaces of fruits and tubers due to suberization.

Proleptic Symptoms

Proleptic symptoms result from the development of tissues earlier than usual. Examples include prolepsis, the premature development of a shoot from a bud, proleptic abscission, the premature formation of abscission layers and restoration, the unexpected development of organs that are normally rudimentary.

Color Changes

Changes in the color of plant tissue are a common symptom of plant disease. Often these color changes are brought about by the yellowing of normal green tissue due to the destruction of chlorophyll or a failure to form chlorophyll. Such repression of leaf color may be complete or partial.

When color repression is complete, it is known as albication. However, the more common, partial repression is referred to as chlorosis.

Patches of green tissue alternating with chlorotic areas are described as a mosaic. Mosaic is a symptom caused by many viruses. Based on the intensity and the pattern of discoloration, mosaics are termed differently. Irregular patches of distinct light and dark areas are known as mottling. Streaking and ring spots are still other distinct types of discolorations. Ring spots are circular masses of chlorosis with a green center. Vein clearing and vein banding are yet other common color changes on leaves.

Chlorophyll may also develop in tissues normally devoid of it. Thus usually white or colored tissue becomes green in color. This is called as virescence.

Anthocyanescence is due to the overdevelopment of anthocyanin and result in the development of a purplish coloration. Color changes can also take place in flowers. Such an example is the color break virus-affected tulips.

Wilts

Wilting is due to loss of turgor in plant tissue resulting in the dropping of plant parts. They are common symptom in diseases where the pathogen or the toxic metabolites it produces affects the vascular tissue of the host plant. Interference in water transport brought about by the infection of these vascular pathogens leads to wilting. Unlike wilting due to low soil moisture, wilting due to the activity of these pathogens cannot be overcome by watering the plants. Infected plants eventually die.

DISEASE RESISTANCE

Plant disease resistance protects plants from pathogens in two ways: by pre-formed structures and chemicals, and by infection-induced responses of the immune system. Relative to a susceptible plant, disease resistance is the reduction of pathogen growth on or in the plant (and hence a reduction of disease), while the term disease tolerance describes plants that exhibit little disease damage despite substantial pathogen levels. Disease outcome is determined by the three-way interaction of the pathogen, the plant and the environmental conditions (an interaction known as the disease triangle).

Defense-activating compounds can move cell-to-cell and systematically through the plant's vascular system. However, plants do not have circulating immune cells, so most cell types exhibit a broad suite of antimicrobial defenses. Although obvious *qualitative* differences in disease resistance can be observed when multiple specimens are compared (allowing classification as "resistant" or "susceptible" after infection by the same pathogen strain at similar inoculum levels in similar environments), a gradation of *quantitative* differences in disease resistance is more typically observed between plant strains or genotypes. Plants consistently resist certain pathogens but succumb to others; resistance is usually specific to certain pathogen species or pathogen strains.

Plant disease resistance is crucial to the reliable production of food, and it provides significant reductions in agricultural use of land, water, fuel and other inputs. Plants in both natural and cultivated populations carry inherent disease resistance, but this has not always protected them.

The late blight Irish potato famine of the 1840s was caused by the oomycete Phytophthora infestans. The world's first mass-cultivated banana cultivar Gros Michel was lost in the 1920s to Panama disease caused by the fungus Fusarium oxysporum. The current wheat stem rust, leaf rust and yellow stripe rust epidemics spreading from East Africa into the Indian subcontinent are caused by rust fungi Puccinia graminis and P. striiformis. Other epidemics include Chestnut blight, as well as recurrent severe plant diseases such as Rice blast, Soybean cyst nematode, Citrus canker.

Plant pathogens can spread rapidly over great distances, vectored by water, wind, insects, and humans. Across large regions and many crop species, it is estimated that diseases typically reduce plant yields by 10% every year in more developed nations or agricultural systems, but yield loss to diseases often exceeds 20% in less developed settings.

However, disease control is reasonably successful for most crops. Disease control is achieved by use of plants that have been bred for good resistance to many diseases, and by plant cultivation approaches such as crop rotation, pathogen-free seed, appropriate planting date and plant density, control of field moisture, and pesticide use.

Common Disease Resistance Mechanisms

Pre-formed Structures and Compounds

- Plant cuticle/surface

- Plant cell walls
- Antimicrobial chemicals (for example: glucosides, saponins)
- Antimicrobial proteins
- Enzyme inhibitors
- Detoxifying enzymes that break down pathogen-derived toxins
- Receptors that perceive pathogen presence and activate inducible plant defences

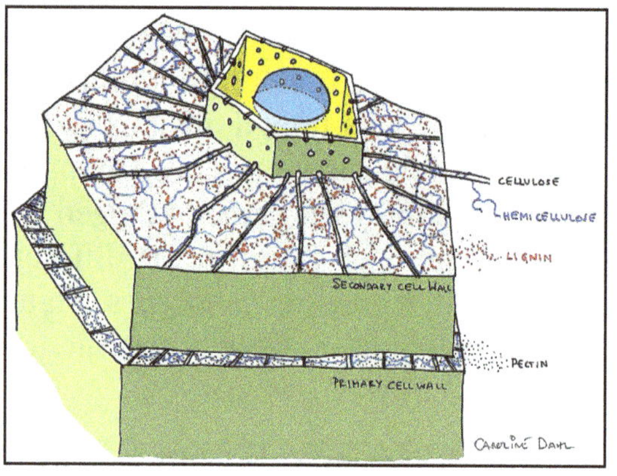

Secondary plant wall.

Inducible Post-Infection Plant Defenses

- Cell wall reinforcement (cellulose, lignin, suberin, cell wall proteins)
- Antimicrobial chemicals, including reactive oxygen species such as hydrogen peroxide or peroxynitrite, or more complex phytoalexins such as genistein or camalexin
- Antimicrobial proteins such as defensins, thionins, or PR-1
- Antimicrobial enzymes such as chitinases, beta-glucanases, or peroxidases
- Hypersensitive response - a rapid host cell death response associated with defence mediated by "Resistance genes.".

Immune System

The plant immune system carries two interconnected tiers of receptors, one most frequently sensing molecules outside the cell and the other most frequently sensing molecules inside the cell. Both systems sense the intruder and respond by activating antimicrobial defenses in the infected cell and neighboring cells. In some cases, defense-activating signals spread to the rest of the plant or even to neighboring plants. The two systems detect different types of pathogen molecules and classes of plant receptor proteins.

The first tier is primarily governed by pattern recognition receptors that are activated by recognition of evolutionarily conserved pathogen or microbial–associated molecular patterns (PAMPs or MAMPs). Activation of PRRs leads to intracellular signaling, transcriptional reprogramming, and biosynthesis of a complex output response that limits colonization. The system is known as PAMP-Triggered Immunity or as Pattern-Triggered Immunity (PTI).

The second tier, primarily governed by R gene products, is often termed effector-triggered immunity (ETI). ETI is typically activated by the presence of specific pathogen "effectors" and then triggers strong antimicrobial responses.

In addition to PTI and ETI, plant defenses can be activated by the sensing of damage-associated compounds (DAMP), such as portions of the plant cell wall released during pathogenic infection.

Responses activated by PTI and ETI receptors include ion channel gating, oxidative burst, cellular redox changes, or protein kinase cascades that directly activate cellular changes (such as cell wall reinforcement or antimicrobial production), or activate changes in gene expression that then elevate other defensive responses.

Plant immune systems show some mechanistic similarities with the immune systems of insects and mammals, but also exhibit many plant-specific characteristics. The two above-described tiers are central to plant immunity but do not fully describe plant immune systems. In addition, many specific examples of apparent PTI or ETI violate common PTI/ETI definitions, suggesting a need for broadened definitions and/or paradigms.

Pattern-Triggered Immunity

PAMPs, conserved molecules that inhabit multiple pathogen genera, are referred to as MAMPs by many researchers. The defenses induced by MAMP perception are sufficient to repel most pathogens. However, pathogen effector proteins are adapted to suppress basal defenses such as PTI. Many receptors for MAMPs (and DAMPs) have been discovered. MAMPs and DAMPs are often detected by transmembrane receptor-kinases that carry LRR or LysM extracellular domains.

Effector Triggered Immunity

Effector Triggered Immunity (ETI) is activated by the presence of pathogen effectors. The ETI response is reliant on R genes, and is activated by specific pathogen strains. Plant ETI often causes an apoptotic hypersensitive response.

R Genes and R Proteins

Plants have evolved R genes (resistance genes) whose products mediate resistance to specific virus, bacteria, oomycete, fungus, nematode or insect strains. R gene products are proteins that allow recognition of specific pathogen effectors, either through direct binding or by recognition of the effector's alteration of a host protein. Many R genes encode NB-LRR proteins (proteins with nucleotide-binding and leucine-rich repeat domains, also known as NLR proteins or STAND proteins, among other names). Most plant immune systems carry a repertoire of 100-600 different R gene homologs. Individual R genes have been demonstrated to mediate resistance to specific virus, bacteria, oomycete, fungus, nematode or insect strains. R gene products control a broad set of disease resistance responses whose induction is often sufficient to stop further pathogen growth/spread.

Studied R genes usually confer specificity for particular strains of a pathogen species (those that express the recognized effector). As first noted by Harold Flor in his mid-20th century formulation of the gene-for-gene relationship, a plant R gene has specificity for a pathogen avirulence gene (Avr gene). Avirulence genes are now known to encode effectors. The pathogen Avr gene must have matched specificity with the R gene for that R gene to confer resistance, suggesting a receptor/ligand interaction for Avr and R genes. Alternatively, an effector can modify its host cellular target (or a molecular decoy of that target), and the R gene product (NLR protein) activates defenses when it detects the modified form of the host target or decoy.

Effector Biology

Effectors are central to the pathogenic or symbiotic potential of microbes and microscopic plant-colonizing animals such as nematodes. Effectors typically are proteins that are delivered outside the microbe and into the host cell. These colonist-derived effectors manipulate the host's cell physiology and development. As such, effectors offer examples of co-evolution (example: a fungal protein that functions outside of the fungus but inside of plant cells has evolved to take on plant-specific functions). Pathogen host range is determined, among other things, by the presence of appropriate effectors that allow colonization of a particular host. Pathogen-derived effectors are a powerful tool to identify plant functions that play key roles in disease and in disease resistance. Apparently most effectors function to manipulate host physiology to allow disease to occur. Well-studied bacterial plant pathogens typically express a few dozen effectors, often delivered into the host by a Type III secretion apparatus. Fungal, oomycete and nematode plant pathogens apparently express a few hundred effectors.

So-called "core" effectors are defined operationally by their wide distribution across the population of a particular pathogen and their substantial contribution to pathogen virulence. Genomics can be used to identify core effectors, which can then be used to discover new R gene alleles, which can be used in plant breeding for disease resistance.

Small RNAs and RNA Interference

Plant sRNA pathways are understood to be important components of pathogen-associated molecular pattern (PAMP)-triggered immunity (PTI) and effector-triggered immunity (ETI). Bacteria-induced miRNAs in Arabidopsis have been shown to influence hormonal signalling including auxin, abscisic acid (ABA), jasmonic acid (JA) and salicylic acid (SA). Advances in genome-wide studies revealed a massive adaptation of host miRNA expression patterns after infection by fungal pathogens *Fusarium virguliforme, Erysiphe graminis, Verticillium dahliae,* and *Cronartium quercuum*, and the oomycete *Phytophthora sojae*. Changes to sRNA expression in response to fungal pathogens indicate that gene silencing may be involved in this defense pathway. However, there is also evidence that the antifungal defense response to *Colletotrichum* spp. infection in maize is not entirely regulated by specific miRNA induction, but may instead act to fine-tune the balance between genetic and metabolic components upon infection.

Transport of sRNAs during infection is likely facilitated by extracellular vesicles (EVs) and multivesicular bodies (MVBs). The composition of RNA in plant EVs has not been fully evaluated, but it is likely that they are, in part, responsible for trafficking RNA. Plants can transport viral RNAs, mRNAs, microRNAs (miRNAs) and small interfering RNAs (siRNAs) systemically through the phloem. This process is thought to occur through the plasmodesmata and involves RNA-binding proteins that assist RNA localization in mesophyll cells. Although they have been identified in the phloem with mRNA, there is no determinate evidence that they mediate long-distant transport of RNAs. EVs may therefore contribute to an alternate pathway of RNA loading into the phloem, or could possibly transport RNA through the apoplast. There is also evidence that plant EVs can allow for interspecies transfer of sRNAs by RNA interference such as Host-Induced Gene Silencing (HIGS). The transport of RNA between plants and fungi seems to be bidirectional as sRNAs from the fungal pathogen *Botrytis cinerea* have been shown to target host defense genes in Arabidopsis and tomato.

Species-Level Resistance

In a small number of cases, plant genes are effective against an entire pathogen species, even though that species that is pathogenic on other genotypes of that host species. Examples include barley MLO against powdery mildew, wheat Lr34 against leaf rust and wheat Yr36 against wheat stripe rust. An array of mechanisms for this type of resistance may exist depending on the particular gene and plant-pathogen combination. Other reasons for effective plant immunity can include a lack of coadaptation (the pathogen and/or plant lack multiple mechanisms needed for colonization and growth within that host species), or a particularly effective suite of pre-formed defenses.

Signaling Mechanisms

Perception of Pathogen Presence

Plant defense signaling is activated by the pathogen-detecting receptors. The activated receptors frequently elicit reactive oxygen and nitric oxide production, calcium, potassium and proton ion fluxes, altered levels of salicylic acid and other hormones and activation of MAP kinases and other specific protein kinases. These events in turn typically lead to the modification of proteins that control gene transcription, and the activation of defense-associated gene expression.

Transcription Factors and the Hormone Response

Numerous genes and/or proteins as well as other molecules have been identified that mediate plant defense signal transduction. Cytoskeleton and vesicle trafficking dynamics help to orient plant defense responses toward the point of pathogen attack.

Mechanisms of Transcription Factors and Hormones

Plant immune system activity is regulated in part by signaling hormones such as:

- Salicylic acid
- Jasmonic acid
- Ethylene

There can be substantial cross-talk among these pathways.

Regulation by Degradation

As with many signal transduction pathways, plant gene expression during immune responses can be regulated by degradation. This often occurs when hormone binding to hormone receptors stimulates ubiquitin-associated degradation of repressor proteins that block expression of certain genes. The net result is hormone-activated gene expression. Examples:

- Auxin: Binds to receptors that then recruit and degrade repressors of transcriptional activators that stimulate auxin-specific gene expression.
- Jasmonic acid: Similar to auxin, except with jasmonate receptors impacting jasmonate-response signaling mediators such as JAZ proteins.
- Gibberellic acid: Gibberellin causes receptor conformational changes and binding and degradation of Della proteins.
- Ethylene: Inhibitory phosphorylation of the EIN2 ethylene response activator is blocked by ethylene binding. When this phosphorylation is reduced, EIN2 protein is cleaved and a portion of the protein moves to the nucleus to activate ethylene-response gene expression.

Ubiquitin and E3 Signaling

Ubiquitination plays a central role in cell signaling that regulates processes including protein degradation and immunological response. Although one of the main functions of ubiquitin is to target proteins for destruction, it is also useful in signaling pathways, hormone release, apoptosis and translocation of materials throughout the cell. Ubiquitination is a component of several immune responses. Without ubiquitin's proper functioning, the invasion of pathogens and other harmful molecules would increase dramatically due to weakened immune defenses.

E3 Signaling

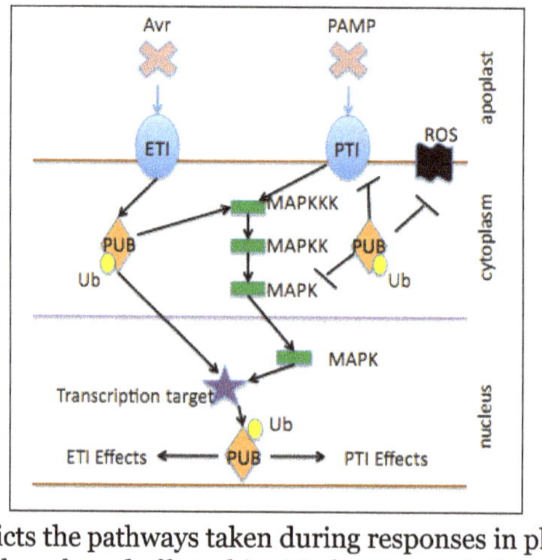

This image depicts the pathways taken during responses in plant immunity. It highlights the role and effect ubiquitin has in regulating the pathway.

The E3 Ubiquitin ligase enzyme is a main component that provides specificity in protein degradation pathways, including immune signaling pathways. The E3 enzyme components can be grouped by which domains they contain and include several types. These include the Ring and U-box single subunit, HECT, and CRLs. Plant signaling pathways including immune responses are controlled by several feedback pathways, which often include negative feedback; and they can be regulated by De-ubiquitination enzymes, degradation of transcription factors and the degradation of negative regulators of transcription.

Plant Breeding for Disease Resistance

Plant breeders emphasize selection and development of disease-resistant plant lines. Plant diseases can also be partially controlled by use of pesticides and by cultivation practices such as crop rotation, tillage, planting density, disease-free seeds and cleaning of equipment, but plant varieties with inherent (genetically determined) disease resistance are generally preferred. Breeding for disease resistance began when plants were first domesticated. Breeding efforts continue because pathogen populations are under selection pressure for increased virulence, new pathogens appear, evolving cultivation practices and changing climate can reduce resistance and/or strengthen pathogens, and plant breeding for other traits can disrupt prior resistance. A plant line with acceptable resistance against one pathogen may lack resistance against others.

Breeding for resistance typically includes:

- Identification of plants that may be less desirable in other ways, but which carry a useful disease resistance trait, including wild plant lines that often express enhanced resistance.
- Crossing of a desirable but disease-susceptible variety to a plant that is a source of resistance.
- Growth of breeding candidates in a disease-conducive setting, possibly including pathogen inoculation. Attention must be paid to the specific pathogen isolates, to address variability within a single pathogen species.
- Selection of disease-resistant individuals that retain other desirable traits such as yield, quality and including other disease resistance traits.

Resistance is termed *durable* if it continues to be effective over multiple years of widespread use as pathogen populations evolve. "Vertical resistance" is specific to certain races or strains of a pathogen species, is often controlled by single R genes and can be less durable. Horizontal or broad-spectrum resistance against an entire pathogen species is often only incompletely effective, but more durable, and is often controlled by many genes that segregate in breeding populations.

Crops such as potato, apple, banana and sugarcane are often propagated by vegetative reproduction to preserve highly desirable plant varieties, because for these species, outcrossing seriously disrupts the preferred traits. Vegetatively propagated crops may be among the best targets for resistance improvement by the biotechnology method of plant transformation to manage genes that affect disease resistance.

Scientific breeding for disease resistance originated with Sir Rowland Biffen, who identified a single recessive gene for resistance to wheat yellow rust. Nearly every crop was then bred to include disease resistance (R) genes, many by introgression from compatible wild relatives.

GM or Transgenic Engineered Disease Resistance

The term GM ("genetically modified") is often used as a synonym of transgenic to refer to plants modified using recombinant DNA technologies. Plants with transgenic/GM disease resistance against insect pests have been extremely successful as commercial products, especially in maize and cotton, and are planted annually on over 20 million hectares in over 20 countries worldwide. Transgenic plant disease resistance against microbial pathogens was first demonstrated in 1986. Expression of viral coat protein gene sequences conferred virus resistance via small RNAs. This proved to be a widely applicable mechanism for inhibiting viral replication. Combining coat protein genes from three different viruses, scientists developed squash hybrids with field-validated, multiviral resistance. Similar levels of resistance to this variety of viruses had not been achieved by conventional breeding.

A similar strategy was deployed to combat papaya ringspot virus, which by 1994 threatened to destroy Hawaii's papaya industry. Field trials demonstrated excellent efficacy and high fruit quality. By 1998 the first transgenic virus-resistant papaya was approved for sale. Disease resistance has been durable for over 15 years. Transgenic papaya accounts for ~85% of Hawaiian production. The fruit is approved for sale in the U.S. Canada and Japan.

Potato lines expressing viral replicase sequences that confer resistance to potato leafroll virus were sold under the trade names NewLeaf Y and NewLeaf Plus, and were widely accepted in commercial production in 1999-2001, until McDonald's Corp. decided not to purchase GM potatoes and Monsanto decided to close their NatureMark potato business. NewLeaf Y and NewLeaf Plus potatoes carried two GM traits, as they also expressed Bt-mediated resistance to Colorado potato beetle.

No other crop with engineered disease resistance against microbial pathogens had reached the market by 2013, although more than a dozen were in some state of development and testing.

Examples of transgenic disease resistance projects				
Publication year	Crop	Disease resistance	Mechanism	Development status
2012	Tomato	Bacterial spot	R gene from pepper	8 years of field trials
2012	Rice	Bacterial blight and bacterial streak	Engineered E gene	Laboratory
2012	Wheat	Powdery mildew	Overexpressed R gene from wheat	2 years of field trials at time of publication
2011	Apple	Apple scab fungus	Thionin gene from barley	4 years of field trials at time of publication
2011	Potato	Potato virus Y	Pathogen-derived resistance	1 year of field trial at time of publication
2010	Apple	Fire blight	Antibacterial protein from moth	12 years of field trials at time of publication
2010	Tomato	Multibacterial resistance	PRR from Arabidopsis	Laboratory scale
2010	Banana	Xanthomonas wilt	Novel gene from pepper	Now in field trial
2009	Potato	Late blight	R genes from wild relatives	3 years of field trials
2009	Potato	Late blight	R gene from wild relative	2 years of field trials at time of publication

2008	Potato	Late blight	R gene from wild relative	2 years of field trials at time of publication
2008	Plum	Plum pox virus	Pathogen-derived resistance	Regulatory approvals, no commercial sales
2005	Rice	Bacterial streak	R gene from maize	Laboratory
2002	Barley	Stem rust	Resting lymphocyte kinase (RLK) gene from resistant barley cultivar	Laboratory
1997	Papaya	Ring spot virus	Pathogen-derived resistance	Approved and commercially sold since 1998, sold into Japan since 2012
1995	Squash	Three mosaic viruses	Pathogen-derived resistance	Approved and commercially sold since 1994
1993	Potato	Potato virus X	Mammalian interferon-induced enzyme	3 years of field trials at time of publication

PRR Transfer

Research aimed at engineered resistance follows multiple strategies. One is to transfer useful PRRs into species that lack them. Identification of functional PRRs and their transfer to a recipient species that lacks an orthologous receptor could provide a general pathway to additional broadened PRR repertoires. For example, the *Arabidopsis* PRR *EF-Tu* receptor (EFR) recognizes the bacterial translation elongation factor *EF-Tu*. Research performed at Sainsbury Laboratory demonstrated that deployment of EFR into either *Nicotiana benthamiana*or *Solanum lycopersicum* (tomato), which cannot recognize *EF-Tu*, conferred resistance to a wide range of bacterial pathogens. EFR expression in tomato was especially effective against the widespread and devastating soil bacterium Ralstonia solanacearum. Conversely, the tomato PRR *Verticillium 1* (*Ve1*) gene can be transferred from tomato to *Arabidopsis*, where it confers resistance to race 1 Verticillium isolates.

Stacking

The second strategy attempts to deploy multiple NLR genes simultaneously, a breeding strategy known as stacking. Cultivars generated by either DNA-assisted molecular breeding or gene transfer will likely display more durable resistance, because pathogens would have to mutate multiple effector genes. DNA sequencing allows researchers to functionally "mine" NLR genes from multiple species/strains.

The *avrBs2* effector gene from *Xanthomona perforans* is the causal agent of bacterial spot disease of pepper and tomato. The first "effector-rationalized" search for a potentially durable R gene followed the finding that *avrBs2* is found in most disease-causing *Xanthomonas* species and is required for pathogen fitness. The *Bs2* NLR gene from the wild pepper, *Capsicum chacoense*, was moved into tomato, where it inhibited pathogen growth. Field trials demonstrated robust resistance without bactericidal chemicals. However, rare strains of *Xanthomonas* overcame *Bs2*-mediated resistance in pepper by acquisition of *avrBs2* mutations that avoid recognition but retain virulence. Stacking R genes that each recognize a different core effector could delay or prevent adaptation.

More than 50 loci in wheat strains confer disease resistance against wheat stem, leaf and yellow stripe rust pathogens. The Stem rust 35 (*Sr35*) NLR gene, cloned from a diploid relative of cultivated wheat, *Triticum monococcum*, provides resistance to wheat rust isolate *Ug99*. Similarly, *Sr33*, from the wheat relative *Aegilops tauschii*, encodes a wheat ortholog to barley *Mla* powdery mildew–resistance genes. Both genes are unusual in wheat and its relatives. Combined with the *Sr2* gene that acts additively with at least Sr33, they could provide durable disease resistance to *Ug99* and its derivatives.

Executor Genes

Another class of plant disease resistance genes opens a "trap door" that quickly kills invaded cells, stopping pathogen proliferation. Xanthomonas and Ralstonia transcription activator–like (TAL) effectors are DNA-binding proteins that activate host gene expression to enhance pathogen virulence. Both the rice and pepper lineages independently evolved TAL-effector binding sites that instead act as an executioner that induces hypersensitive host cell death when up-regulated. *Xa27* from rice and Bs3 and Bs4c from pepper, are such "executor" (or "executioner") genes that encode non-homologous plant proteins of unknown function. Executor genes are expressed only in the presence of a specific TAL effector.

Engineered executor genes were demonstrated by successfully redesigning the pepper *Bs3* promoter to contain two additional binding sites for TAL effectors from disparate pathogen strains. Subsequently, an engineered executor gene was deployed in rice by adding five TAL effector binding sites to the *Xa27* promoter. The synthetic *Xa27* construct conferred resistance against Xanthomonas bacterial blight and bacterial leaf streak species.

Host Susceptibility Alleles

Most plant pathogens reprogram host gene expression patterns to directly benefit the pathogen. Reprogrammed genes required for pathogen survival and proliferation can be thought of as "disease-susceptibility genes." Recessive resistance genes are disease-susceptibility candidates. For example, a mutation disabled an *Arabidopsis* gene encoding pectate lyase (involved in cell wall degradation), conferring resistance to the powdery mildew pathogen *Golovinomyces cichoracearum*. Similarly, the Barley *MLO* gene and spontaneously mutated pea and tomato *MLO* orthologs also confer powdery mildew resistance.

Lr34 is a gene that provides partial resistance to leaf and yellow rusts and powdery mildew in wheat. *Lr34* encodes an adenosine triphosphate (ATP)–binding cassette (ABC) transporter. The dominant allele that provides disease resistance was recently found in cultivated wheat (not in wild strains) and, like *MLO* provides broad-spectrum resistance in barley.

Natural alleles of host translation elongation initiation factors *eif4e* and *eif4g* are also recessive viral-resistance genes. Some have been deployed to control potyviruses in barley, rice, tomato, pepper, pea, lettuce and melon. The discovery prompted a successful mutant screen for chemically induced *eif4e* alleles in tomato.

Natural promoter variation can lead to the evolution of recessive disease-resistance alleles. For example, the recessive resistance gene *xa13* in rice is an allele of *Os-8N3*. *Os-8N3* is transcriptionally activated by *Xanthomonas oryzae pv. oryzae* strains that express the TAL effector *PthXo1*. The *xa13* gene has a mutated effector-binding element in its promoter that eliminates *PthXo1* binding and renders these lines resistant to strains that rely on *PthXo1*. This finding also demonstrated that *Os-8N3* is required for susceptibility.

Xa13/Os-8N3 is required for pollen development, showing that such mutant alleles can be problematic should the disease-susceptibility phenotype alter function in other processes. However, mutations in the *Os11N3* (OsSWEET14) TAL effector–binding element were made by fusing TAL effectors to nucleases (TALENs). Genome-edited rice plants with altered *Os11N3* binding sites remained resistant to *Xanthomonas oryzae pv. oryzae*, but still provided normal development function.

Gene Silencing

RNA silencing-based resistance is a powerful tool for engineering resistant crops. The advantage of RNAi as a novel gene therapy against fungal, viral and bacterial infection in plants lies in the fact that it regulates gene expression via messenger RNA degradation, translation repression and chromatin remodelling through small non-coding RNAs. Mechanistically, the silencing processes are guided by processing products of the double-stranded RNA (dsRNA) trigger, which are known as small interfering RNAs and microRNAs.

Host Range

Among the thousands of species of plant pathogenic microorganisms, only a small minority have the capacity to infect a broad range of plant species. Most pathogens instead exhibit a high degree of host-specificity. Non-host plant species are often said to express *non-host resistance*. The term *host resistance* is used when a pathogen species can be pathogenic on the host species but certain strains of that plant species resist certain strains of the pathogen species. The causes of host resistance and non-host resistance can overlap. Pathogen host range is determined, among other things, by the presence of appropriate effectors that allow colonization of a particular host. Pathogen host range can change quite suddenly if, for example, the pathogen's capacity to synthesize a host-specific toxin or effector is gained by gene shuffling/mutation, or by horizontal gene transfer.

Epidemics and Population Biology

Native populations are often characterized by substantial genotype diversity and dispersed populations (growth in a mixture with many other plant species). They also have undergone of plant-pathogen coevolution. Hence as long as novel pathogens are not introduced/do not evolve, such populations generally exhibit only a low incidence of severe disease epidemics.

Monocrop agricultural systems provide an ideal environment for pathogen evolution, because they offer a high density of target specimens with similar/identical genotypes. The rise in mobility stemming from modern transportation systems provides pathogens with access to more potential targets. Climate change can alter the viable geographic range of pathogen species and cause some diseases to become a problem in areas where the disease was previously less important.

These factors make modern agriculture more prone to disease epidemics. Common solutions include constant breeding for disease resistance, use of pesticides, use of border inspections and plant import restrictions, maintenance of significant genetic diversity within the crop gene pool, and constant surveillance to accelerate initiation of appropriate responses. Some pathogen species have much greater capacity to overcome plant disease resistance than others, often because of their ability to evolve rapidly and to disperse broadly.

INTEGRATED PEST MANAGEMENT

An IPM boll weevil trap in a cotton field.

Integrated pest management (IPM), also known as integrated pest control (IPC) is a broad-based approach that integrates practices for economic control of pests. IPM aims to suppress pest populations below the economic injury level (EIL). The UN's Food and Agriculture Organization defines IPM as "the careful consideration of all available pest control techniques and subsequent integration of appropriate measures that discourage the development of pest populations and keep pesticides and other interventions to levels that are economically justified and reduce or minimize risks to human health and the environment. IPM emphasizes the growth of a healthy crop with the least possible disruption to agro-ecosystems and encourages natural pest control mechanisms." Entomologists and ecologists have urged the adoption of IPM pest control since the 1970s. IPM allows for safer pest control.

The introduction and spread of invasive species can also be managed with IPM by reducing risks while maximizing benefits and reducing costs.

Principles

An American IPM system is designed around six basic components:

- Acceptable pest levels—The emphasis is on control, not eradication. IPM holds that wiping out an entire pest population is often impossible, and the attempt can be expensive and unsafe. IPM programmes first work to establish acceptable pest levels, called action thresholds, and apply controls if those thresholds are crossed. These thresholds are pest and site specific, meaning that it may be acceptable at one site to have a weed such as white clover, but not at another site. Allowing a pest population to survive at a reasonable threshold reduces selection pressure. This lowers the rate at which a pest develops resistance to a control, because if almost all pests are killed then those that have resistance will provide the genetic basis of the future population. Retaining a significant number of unresistant specimens dilutes the prevalence of any resistant genes that appear. Similarly, the repeated use of a single class of controls will create pest populations that are more resistant to that class, whereas alternating among classes helps prevent this.
- Preventive cultural practices—Selecting varieties best for local growing conditions and maintaining healthy crops is the first line of defense. Plant quarantine and 'cultural techniques' such as crop sanitation are next, e.g. removal of diseased plants, and cleaning

pruning shears to prevent spread of infections. Beneficial fungi and bacteria are added to the potting media of horticultural crops vulnerable to root diseases, greatly reducing the need for fungicides.

- Monitoring—Regular observation is critically important. Observation is broken into inspection and identification. Visual inspection, insect and spore traps, and other methods are used to monitor pest levels. Record-keeping is essential, as is a thorough knowledge of target pest behavior and reproductive cycles. Since insects are cold-blooded, their physical development is dependent on area temperatures. Many insects have had their development cycles modeled in terms of degree-days. The degree days of an environment determines the optimal time for a specific insect outbreak. Plant pathogens follow similar patterns of response to weather and season.
- Mechanical controls—Should a pest reach an unacceptable level, mechanical methods are the first options. They include simple hand-picking, barriers, traps, vacuuming and tillage to disrupt breeding.
- Biological controls—Natural biological processes and materials can provide control, with acceptable environmental impact, and often at lower cost. The main approach is to promote beneficial insects that eat or parasitize target pests. Biological insecticides, derived from naturally occurring microorganisms (e.g.—Bt, entomopathogenic fungi and entomopathogenic nematodes), also fall in this category. Further 'biology-based' or 'ecological' techniques are under evaluation.
- Responsible use—Synthetic pesticides are used as required and often only at specific times in a pest's life cycle. Many newer pesticides are derived from plants or naturally occurring substances (e.g.—nicotine, pyrethrum and insect juvenile hormone analogues), but the toxophore or active component may be altered to provide increased biological activity or stability. Applications of pesticides must reach their intended targets. Matching the application technique to the crop, the pest, and the pesticide is critical. The use of low-volume spray equipment reduces overall pesticide use and labor cost.

An IPM regime can be simple or sophisticated. Historically, the main focus of IPM programmes was on agricultural insect pests. Although originally developed for agricultural pest management, IPM programmes are now developed to encompass diseases, weeds and other pests that interfere with management objectives for sites such as residential and commercial structures, lawn and turf areas, and home and community gardens.

Process

IPM is the selection and use of pest control actions that will ensure favourable economic condition, ecological and social consequences and is applicable to most agricultural, public health and amenity pest management situations. The IPM process starts with monitoring, which includes inspection and identification, followed by the establishment of economic injury levels. The economic injury levels set the economic threshold level. That is the point when pest damage (and the benefits of treating the pest) exceed the cost of treatment. This can also be an action threshold level for determining an unacceptable level that is not tied to economic injury. Action thresholds are more common in structural pest management and economic injury levels in classic agricultural pest management. An example of an action threshold is one fly in a hospital operating room is not acceptable, but one fly in a pet kennel would be acceptable. Once a threshold has been crossed by the pest population action steps need to be taken to reduce and control the pest. Integrated pest management employs a variety of actions including cultural controls such as physical barriers, biological controls such as adding and conserving natural predators and enemies of the pest, and finally chemical controls or pesticides. Reliance on knowledge, experience, observation and integration of multiple techniques makes IPM appropriate for organic farming (excluding synthetic pesticides). These may or may not include materials listed on the Organic Materials Review Institute (OMRI). Although the pesticides and particularly insecticides used in organic farming and organic gardening are generally safer than synthetic pesticides, they are not always more safe or environmentally friendly than synthetic pesticides and can cause harm. For conventional farms IPM can reduce human and environmental exposure to hazardous chemicals, and potentially lower overall costs.

Risk assessment usually includes four issues: Characterization of biological control agents, health risks, environmental risks and efficacy.

Mistaken identification of a pest may result in ineffective actions. E.g. plant damage due to over-watering could be mistaken for fungal infection, since many fungal and viral infections arise under moist conditions.

Monitoring begins immediately, before the pest's activity becomes significant. Monitoring of agricultural pests includes tracking soil/planting media fertility and water quality. Overall plant health and resistance to pests is greatly influenced by pH, alkalinity, of dissolved mineral and oxygen reduction potential. Many diseases are waterborne, spread directly by irrigation water and indirectly by splashing.

Once the pest is known, knowledge of its lifecycle provides the optimal intervention points. For example, weeds reproducing from last year's seed can be prevented with mulches and pre-emergent herbicide.

Pest-tolerant crops such as soybeans may not warrant interventions unless the pests are numerous or rapidly increasing. Intervention is warranted if the expected cost of damage by the pest is more than the cost of control. Health hazards may require intervention that is not warranted by economic considerations.

Specific sites may also have varying requirements. E.g. white clover may be acceptable on the sides of a tee box on a golf course, but unacceptable in the fairway where it could confuse the field of play.

Possible interventions include mechanical/physical, cultural, biological and chemical. Mechanical/physical controls include picking pests off plants, or using netting or other material to exclude pests such as birds from grapes or rodents from structures. Cultural controls include keeping an area free of conducive conditions by removing waste or diseased plants, flooding, sanding, and the use of disease-resistant crop varieties. Biological controls are numerous. They include: conservation of natural predators or augmentation of natural predators, sterile insect technique (SIT).

Augmentation, inoculative release and inundative release are different methods of biological control that affect the target pest in different ways. Augmentative control includes the periodic introduction of predators. With inundative release, predators are collected, mass-reared and periodically released in large numbers into the pest area. This is used for an immediate reduction in host populations, generally for annual crops, but is not suitable for long run use. With inoculative release a limited number of beneficial organisms are introduced at the start of the growing season. This strategy offers long term control as the organism's progeny affect pest populations throughout the season and is common in orchards. With seasonal inoculative release the beneficials are collected, mass-reared and released seasonally to maintain the beneficial population. This is commonly used in greenhouses. In America and other western countries, inundative releases are predominant, while Asia and the eastern Europe more commonly use inoculation and occasional introductions.

The sterile insect technique (SIT) is an area-wide IPM program that introduces sterile male pests into the pest population to trick females into (unsuccessful) breeding encounters, providing a form of birth control and reducing reproduction rates. The biological controls mentioned above only appropriate in extreme cases, because in the introduction of new species, or supplementation of naturally occurring species can have detrimental ecosystem effects. Biological controls can be used to stop invasive species or pests, but they can become an introduction path for new pests.

Chemical controls include horticultural oils or the application of insecticides and herbicides. A green pest management IPM program uses pesticides derived from plants, such as botanicals, or other naturally occurring materials.

Pesticides can be classified by their modes of action. Rotating among materials with different modes of action minimizes pest resistance. Evaluation is the process of assessing whether the intervention was effective, whether it produced unacceptable side effects, whether to continue, revise or abandon the program.

DISEASE CONTROL

Traditional principles of plant disease control:

1. Avoidance—Prevent disease by selecting a time of the year or a site where there is no inoculum or where the environment is not favorable for infection.
2. Exclusion—Prevent the introduction of inoculum.
3. Eradication—Eliminate, destroy, or inactivate the inoculum.
4. Protection—Prevent infection by means of a toxicant or some other barrier to infection.
5. Resistance—Utilize cultivars that are resistant to or tolerant of infection.
6. Therapy—Cure plants that are already infected.

While these principles are as valid today as they were in 1929, in the context of modern concepts of plant disease management, they have some critical shortcomings. First of all, these principles are stated in absolute terms (e.g. "exclude", "prevent", and "eliminate") that imply a goal of zero disease. Plant disease "control" in this sense is not practical, and in most cases is not even possible. Indeed, we need not eliminate a disease; we merely need to reduce its progress and keep disease development below an acceptable level. Instead of plant disease control, we need to think in terms of plant disease management.

A second shortcoming is that the traditional principles of plant disease control do not take into consideration the dynamics of plant disease, that is, the changes in the incidence and severity of disease in time and space. Furthermore, considering that different diseases differ in their dynamics, they do not indicate the relative effectiveness of the various tactics for the control of a particular disease. They also fail to show how the different disease control measures interact in their effects on disease dynamics. We need some means of assessing quantitatively the effects of various control measures, singly and in combination, on the progress of disease.

Finally, the traditional principles of plant disease control tend to emphasize tactics without fitting them into an adequate overall strategy.

Strategies versus Tactics

Ask a handful of pest management experts to name the major plant disease control strategies, and you are sure to find disagreement. The problem is generally one of semantics rather than of fundamental disagreement over the important means of disease control. The dictionary definitions for the two terms are similar, but generally speaking, an overall plan for reaching a particular objective is called a strategy, while the specific means for implementing a given strategy are called tactics. Like the goals and objectives that they are intended to achieve, strategies and tactics tend to occur in hierarchies. What is a "strategy" at one level of focus could be called a "tactic" at another level of focus.

The important point to remember is that countless human undertakings, be they military operations, political campaigns, football games, or any other kind of organized effort, have failed, despite flawless tactics, for lack of a sound strategy. Any endeavor that requires a series of connected tasks for its completion also requires some kind of overall plan. Each individual task, no matter how skillfully executed or how successful its outcome, will not advance progress toward the final objective unless it has a coherent relationship with all of the other necessary tasks.

Epidemiological basis of Disease Management

Plant disease epidemics can be classified into two basic types, monocyclic and polycyclic, depending on the number of infection cycles per crop cycle. The early stages of a monocyclic epidemic can be described quite well by a linear model, while the early stages of a polycyclic epidemic can be described with an exponential model. Since we are concerned with keeping disease levels well below 100%, there is no need to adjust the models for approaching the upper limit, and we can use the simple linear and exponential models to plan strategies:

Monocyclic Model

$$x = \overset{\overset{1}{\downarrow}}{Q}\,\overset{\overset{2}{\downarrow}}{R}\,\overset{\overset{3}{\downarrow}}{t}$$

It is clear from the above model of a monocyclic epidemic that Q, R, and t have equal weight in their effect on x. A reduction in the initial inoculum or the rate of infection will result in a reduction in the level of disease by the same proportion at any time, t, throughout the epidemic. If t can be reduced (for example, by shortening the season), disease will be reduced proportionately.

The Polycyclic Model

$$x = \overset{\overset{1}{\downarrow}}{x_0}\, e^{\overset{\overset{2}{\downarrow}}{r}\,\overset{\overset{3}{\downarrow}}{t}}$$

Examining these models, we can see that in both there are three ways in which we can reduce x at any point in the epidemic:

- Reduce the initial inoculum (Q in the monocyclic model and x_0 in the polycyclic model). (Actually x_0 is the initial incidence of disease, which is proportional to the initial inoculum.)
- Reduce the rate of infection (R in the monocyclic model and r in the polycyclic model).
- Reduce the duration of the epidemic (the time, t, at the end of the epidemic).

These, then, can be used as three major strategies for managing plant disease epidemics, and we can organize our plant disease control tactics under one or more of these overall strategies. Furthermore, by means of the model we can assess the quantitative impact of each strategy, not only by itself, but in its interaction with others.

- If r is very high, the apparent effect of reducing xo is to delay the epidemic.
- If r is very high, x_0 must be reduced to very low levels to have a significant effect on the epidemic.
- Reducing r has a relatively greater effect on the epidemic than reducing x_0.
- Reducing x_0 makes good strategic sense only if r is low or if r is also being reduced.

It is easier to understand these concepts if we actually select different values for x_0 and r, plug them into the model, and graph the outcome. This can be done easily with a calculator that has an exponential function, or with the accompanying simulation.

Clearly developing a sound disease management strategy requires enough knowledge of the biology of the pathogen and host to select the appropriate epidemiological model. It also requires at least "ball-park" estimates of the model parameters and the magnitude of the impact of each specific tactic on the initial inoculum or the apparent infection rate. Failure to adopt such a quantitative approach can lead to some embarrassing or even very costly errors.

Traditional Principles

To make the conceptual leap from disease control to disease management, the traditional principles can be modified by fitting them as tactics within each of the three major disease management strategies and by slightly changing the wording to reflect the quantitative impact of the action rather than an absolute effect:

Tactics for the Reduction of Initial Inoculum

- Avoidance—Reduce the level of disease by selecting a season or a site where the amount of inoculum is low or where the environment is unfavorable for infection.
- Exclusion—Reduce the amount of initial inoculum introduced from outside sources.
- Eradication—Reduce the production of initial inoculum by destroying or inactivating the sources of initial inoculum (sanitation, removal of reservoirs of inoculum, removal of alternate hosts, etc.).
- Protection—Reduce the level of initial infection by means of a toxicant or other barrier to infection.
- Resistance—Use cultivars that are resistant to infection, particularly the initial infection.
- Therapy—Use thermotherapy, chemotherapy and/or meristem culture to produce certified seed or vegetative planting stock.

Tactics for the Reduction of the Infection Rate

- Avoidance—Reduce the rate of production of inoculum, the rate of infection, or the rate of development of the pathogen by selecting a season or a site where the environment is not favorable.
- Exclusion—Reduce the introduction of inoculum from external sources during the course of the epidemic.
- Eradication—Reduce the rate of inoculum production during the course of the epidemic by destroying or inactivating the sources of inoculum (roguing).
- Protection—Reduce the rate of infection by means of a toxicant or some other barrier to infection.
- Resistance—Plant cultivars that can reduce the rate of inoculum production, the rate of infection, or the rate of pathogen development.
- Therapy—Cure the plants that are already infected or reduce their production of inoculum.

Tactics for the Reduction of the Duration of the Epidemic

- Avoidance—Plant early maturing cultivars or plant at a time that favors rapid maturation of the crop.
- Exclusion—Delay the introduction of inoculum from external sources by means of plant quarantine.

References

- Department of Plant Pathology and Environmental Microbiology (Penn State University)". Department of Plant Pathology and Environmental Microbiology (Penn State University). Retrieved 12 December 2018
- Plant Disease Diagnosis, case studies, disimpactmngmnt, edcenter: apsnet.org, Retrieved 13 May, 2019
- Jones, J. D.; Dangl, J.L. (2006). "The plant immune system". Nature. 444 (7117): 323–329. Bibcode:2006Natur.444..323J. Doi:10.1038/nature05286. PMID 17108957
- IPM Guidelines". Umassamherst—Integrated Pest Management, Agriculture and Landscape Program. 2009. Archived from the original on 12 March 2012. Retrieved 13 March 2012
- J. C. Van Lenteren (2003). Quality Control and Production of Biological Control Agents: Theory and Testing Procedures. CABI. ISBN 978-0-85199-836-7
- Witches' Brooms on Trees - Horticulture and Home Pest News". Hortnews.extension.iastate.edu. Retrieved 12 December 2018
- Management Strategies, Epidemiology Temporal, disimpactmngmnt, apsnet.org, Retrieved 25 February, 2019

Permissions

All chapters in this book are published with permission under the Creative Commons Attribution Share Alike License or equivalent. Every chapter published in this book has been scrutinized by our experts. Their significance has been extensively debated. The topics covered herein carry significant information for a comprehensive understanding. They may even be implemented as practical applications or may be referred to as a beginning point for further studies.

We would like to thank the editorial team for lending their expertise to make the book truly unique. They have played a crucial role in the development of this book. Without their invaluable contributions this book wouldn't have been possible. They have made vital efforts to compile up to date information on the varied aspects of this subject to make this book a valuable addition to the collection of many professionals and students.

This book was conceptualized with the vision of imparting up-to-date and integrated information in this field. To ensure the same, a matchless editorial board was set up. Every individual on the board went through rigorous rounds of assessment to prove their worth. After which they invested a large part of their time researching and compiling the most relevant data for our readers.

The editorial board has been involved in producing this book since its inception. They have spent rigorous hours researching and exploring the diverse topics which have resulted in the successful publishing of this book. They have passed on their knowledge of decades through this book. To expedite this challenging task, the publisher supported the team at every step. A small team of assistant editors was also appointed to further simplify the editing procedure and attain best results for the readers.

Apart from the editorial board, the designing team has also invested a significant amount of their time in understanding the subject and creating the most relevant covers. They scrutinized every image to scout for the most suitable representation of the subject and create an appropriate cover for the book.

The publishing team has been an ardent support to the editorial, designing and production team. Their endless efforts to recruit the best for this project, has resulted in the accomplishment of this book. They are a veteran in the field of academics and their pool of knowledge is as vast as their experience in printing. Their expertise and guidance has proved useful at every step. Their uncompromising quality standards have made this book an exceptional effort. Their encouragement from time to time has been an inspiration for everyone.

The publisher and the editorial board hope that this book will prove to be a valuable piece of knowledge for students, practitioners and scholars across the globe.

Index

www.ingramcontent.com/pod-product-compliance
Lightning Source LLC
LaVergne TN
LVHW080647210926
848485LV00003B/341

* 9 7 8 1 6 4 7 4 0 0 6 5 1 *